The Statue Within

BOOKS IN THE ALFRED P. SLOAN FOUNDATION SERIES

Disturbing the Universe *by Freeman Dyson*
Advice to a Young Scientist *by Peter Medawar*
The Youngest Science *by Lewis Thomas*
Haphazard Reality *by Hendrik B. G. Casimir*
In Search of Mind *by Jerome Bruner*
A Slot Machine, a Broken Test Tube *by S. E. Luria*
Enigmas of Chance *by Mark Kac*
Rabi: Scientist and Citizen *by John Rigden*
Alvarez: Adventures of a Physicist *by Luis W. Alvarez*
Making Weapons, Talking Peace *by Herbert F. York*
In Praise of Imperfection *by Rita Levi-Montalcini*
Memoirs of an Unregulated Economist *by George J. Stigler*
What Mad Pursuit *by Francis Crick*

THIS BOOK IS PUBLISHED AS PART OF AN ALFRED P. SLOAN FOUNDATION PROGRAM

THE STATUE

WITHIN

An Autobiography

FRANÇOIS JACOB

Translated by Franklin Philip

Basic Books, Inc., Publishers   New York

Library of Congress Cataloging-in-Publication Data

Jacob, François, 1920–
    The statue within.

    (Alfred P. Sloan Foundation series)
    Translation of: La statue intérieure.
    Includes index.
    1. Jacob, François, 1920– . 2. Geneticists—
France—Biography. 3. Molecular biologists—France
—Biography. I. Title. II. Series.
QH429.2.J33A3   1988   575.1'092'4[B]   87–47780
ISBN 0–465–08223–8 (cloth)
ISBN 0–465–08222–X (paper)

Preface to the Series

THE ALFRED P. SLOAN FOUNDATION has for many years had an interest in encouraging public understanding of science. Science in this century has become a complex endeavor. Scientific statements may reflect many centuries of experimentation and theory, and are likely to be expressed in the language of advanced mathematics or in highly technical terms. As scientific knowledge expands, the goal of general public understanding of science becomes increasingly difficult to reach.

Yet an understanding of the scientific enterprise, as distinct from data, concepts, and theories, is certainly within the grasp of us all. It is an enterprise conducted by men and women who are stimulated by hopes and purposes that are universal, rewarded by occasional successes, and distressed by setbacks. Science is an enterprise with its own rules and customs, but an understanding of that enterprise is accessible, for it is quintessentially human. And an understanding of the enterprise inevitably brings with it insights into the nature of its products.

The Sloan Foundation expresses great appreciation to the advisory committee. Present members include the chairman, Simon Michael Bessie, Co-Publisher, Cornelia and Michael Bessie Books; Howard Hiatt, Professor, School of Medicine, Harvard

Preface to the Series

University; Eric R. Kandel, University Professor, Columbia University College of Physicians and Surgeons, and Senior Investigator, Howard Hughes Medical Institute; Daniel Kevles, Professor of History, California Institute of Technology; Robert Merton, University Professor Emeritus, Columbia University; Paul Samuelson, Institute Professor of Economics, Massachusetts Institute of Technology; Robert Sinsheimer, Chancellor Emeritus, University of California, Santa Cruz; Steven Weinberg, Professor of Physics, University of Texas at Austin; and Stephen White, former Vice-President of the Alfred P. Sloan Foundation. Previous members of the committee were Daniel McFadden, Professor of Economics, and Philip Morrison, Professor of Physics, both of the Massachusetts Institute of Technology; Mark Kac (deceased), formerly Professor of Mathematics, University of Southern California; George Miller, Professor of Psychology, Princeton University; and Frederick E. Terman (deceased), formerly Provost Emeritus, Stanford University. The Sloan Foundation has been represented by Arthur L. Singer, Jr., Stephen White, Eric Wanner, and Sandra Panem. The first publisher of the program, Harper & Row, was represented by Edward L. Burlingame and Sallie Coolidge. This volume is the fourth to be published by Basic Books, represented by Martin Kessler and Richard Liebmann-Smith.

—ALBERT REES
President
Alfred P. Sloan Foundation

The Statue Within

I

LAST NIGHT I woke up haunted by the memory of a friend now dead. Jean S. An old wartime pal. He had been involved in every Free French campaign until, in a forest in Normandy, a burst of heavy machinegun fire hit him in the thigh. The leg was amputated. Excruciating pains in the phantom limb. In the hospital for months on end. Every possible treatment, every sort of medication, every painkiller. Under narcotic sedation, he was asked to relive the day he had been wounded. He recalled entering the forest; advancing step by step, tree by tree; the Germans' fierce counterattack. And suddenly the violent blow in his thigh. When he screamed from the shock of the recollection, he was awakened and assured the pain was gone. It was still there, of course, as imperious, as compelling as ever. Later, he was to be seen stumping along on an artificial leg, with his slightly embittered smile, warm but often constricted in a grimace of unremitting pain. The most affecting thing about him had been his vitality and capacity for hopefulness. But the last time he came to see me at the Pasteur Institute, I noticed a change in him: more voluble speech, more feverish gesturing, an anxious look in his eyes. He wanted a referral to a medical specialist: a paralysis in the pelvic area, the bladder, and rectum threatened to add to his woes. As he talked, however, I began to hear another message, to detect the anguish of a hidden question. In the flood of words was surfacing another request, a plea for help. What he was seeking from me was a guarantee against further disaster, a promise to relieve him, to stop the debacle, to help him go when the time came. But this secret prayer I would not hear. Cowardly, I ignored it.

Words sometimes have odd associations. For me, the word *suicide* was long associated with two illustrations from books about antiquity. When I was seven or eight years old, my grandfather the general, who had taken over my instruction in classical culture, wanted me to find more amusement in the tales and legends of ancient history than in comic strips. So he fed me books about the pharaohs, the founding of Rome, Greek mythology, books that I read again and again. To this day I can still see some of the illustrations in exact detail. The death of Cleopatra, for example. The queen is seated on a stool at a table where a kneeling slave is setting a basket of fruit. Wearing the royal crown, naked to the waist, hieratic, Cleopatra holds out an arm to the asp rising, hissing, from amid the figs. Strangely, for a long time what the word *suicide* connoted for me was this woman's authority in death, her beauty. It was the powerful shoulder, the lithe extended arm, the shape of the proud and naked breast. The other illustration showed Socrates in prison, meditating and holding in his hand the goblet of hemlock. I couldn't understand this situation. Forcing a man to kill himself instead of executing him, using poison rather than the sword—all this seemed unfair and out of keeping with the ideas I had formed about the Greeks. The word *suicide*, in its very sound and its written form, was associated with these two illustrations: the first syllable—*sui*—writhing and hissing like a snake, especially the asp, owing to the long *e* sound of *i* in French; and the final *cide*, pronounced like the English *seed*, as acid and deadly as the hemlock. Thus, in death as self-punishment, Cleopatra came to be allied with Socrates: the body and the head, the smooth and the hairy, sex and thought, power and intellect. Cleopatra was, moreover, only one among many such examples. During that period of my life, I spent many hours in the company of Greek goddesses and Roman ladies. They aroused in me many emotions. While, in picture books and statues in museums, I admired these women's bodies, lacking constraints, nude or in drapery, hard and wonderfully polished, I was also learning that these same proud women were sometimes buried alive, or did not shrink from having their slaves thrown to sea lampreys. All this provided much material for dreaming. Greek goddesses and Roman ladies have, in fact, played a large role in my life.

Death, extinction, "passing away," as they say, is best forgotten. In everyday life, in physical behavior, it is impossible to live as if under sentence of death. Although one is not responsible for one's birth, one is in some way responsible for one's death. What one cannot forget is the fear of being afraid; the distaste for becoming distasteful; the inability to escape impotence. And also the terror of being dominated like a child, of being manipulated; the obsession with becoming other than one is, with thinking differently, and even with thinking no more at all. Then the nightmare of having to submit, of being acted upon, unable to react, to explain oneself, or even to make a request. In short, the specter of becoming a vegetable. At this point, Cleopatra and Socrates reappear. At this point, the poison loses its unfair character and becomes an ally, as it was in the Resistance during the war, to keep from talking under torture. The hard part is choosing the right moment. Too soon is stupid; too late is not possible. And to consider these alternatives is perhaps already to avoid the issue. In this matter, there is no ideal moment.

My grandmother, my mother's mother, the general's wife, died at the age of ninety-seven. A splendid, impressive woman, she long remained a sort of Junoesque presence. Yet she had a difficult personality, headstrong, imbued with the certainty of always being right. During the war, she was arrested in Lyon under a false name, and taken to the Gestapo, who interrogated her to make her confess that she was Jewish and tell them the hiding place of her Jewish son, whom they suspected of working for the Resistance. For a whole afternoon in the August heat, with the German police questioning her in relays, she confronted them head-on, loftily refusing to answer the Germans' questions unless they showed her the respect she judged to be due the widow of a French general. She even managed to become genuinely angry, bawling out the German police so forcefully that, dazed, they released her. Then she vanished into thin air until the Occupation was over. Imperious and high-handed, she never in all her life stopped trying to run the whole show, and that included her husband, the general. She meant to be clear-headed, as much about herself as about others. "I don't want to get old," she would say. "I don't want to be a burden to anyone. Being diminished, ugh! I won't put up with it! I'll know how to get out in time." Up to the age of ninety-five, she remained "*la générale*"—straight as the letter I, chignon tightly coiled, a spotless white band around her neck.

Though past fifty then, I was still her "little one." Shortly thereafter, and very quickly, came the collapse. The last time I went to see her in Lyon, she was no longer leaving her room. Seeing me come in, she sat up in bed. "Why don't you want to take me to lunch at that little restaurant near the military college where you get such good stuffed cabbage for five *sous*?" It took me a while to realize that she had regressed to a time before the First World War and was taking me for her brother, dead more than ten years. Dumbfounded and with a lump in my throat, I gazed at this old woman in whose house I had often spent my school vacations; who played with me for hours on end; taught me the names of flowers and bugs; caught butterflies and found birds' nests for me in the gardens of the Headquarters in Le Mans, where my grandfather was then in command of the army corps. On her night table, in the gallery of those she had loved, was a photo of me as a baby in her arms. Now, in front of this poor old woman adrift in time and in space, her identity lost, I stood frozen, incapable of a gesture or a word. After a long silence, I went slowly up to her to embrace her warmly, to give her a kiss. She pushed me back, furious: "First give me back those marbles you took from me." Her brother had been a doctor—as was I, though not a practicing one. Perhaps, in this violence toward her brother, whom that day I embodied for her, there was a terrible reproach for not having been helped to avoid this disaster; to "get out in time." Perhaps, alongside the muddled images stirred up by the past, she was accusing us as doctors of failing to come to the aid of someone in distress. Or perhaps she merely wanted to let us know we had been unable to love her as she wished.

My obsession: a life that shrivels up, slowly rots, goes soft as a pulp. This worry about decline grabs me by the throat as I awake. In the brief interval between dream and waking, it flaunts before my eyes the frenzied dance of everything I would have liked to do, and did not do, and never will. As I turn over and over in my bed, the fear of the too-late, of the irreversible, propels me to the mirror to shave and get ready for the day. And that is the moment of truth. The moment for the old questions. What am I today? Am I capable of renewal? What are the chances I might still produce something I do not expect of myself? For my life unfolds mainly in the yet-to-come, and is based on waiting. Mine is a life of preparation. I enjoy the present only insofar as it is a promise of the future. I am looking for the Promised Land and listening to the music of my tomorrows. My food is anticipation. My drug is hope. As a child, unable to bear the absence of a goal, I made out of trifles what I called "little lights" to illuminate the coming day or week. In writing this memoir, I aim neither to wallow in the mire of self-satisfaction nor to settle old scores, but rather to set myself a new purpose, and thus a new existence. It is to take my past and produce the future. I am bored by what has been done, and excited only by what is to do. Were I to frame a prayer, I would ask to be granted not so much the "strength" as the "desire" to do.

This endless race with time, this preference for desire over enjoyment is not without its drawbacks. Too often, it prevents us from understanding, and nurtures the illusion of life rather than life itself. It took me a long time to realize that this drive toward tomorrow has an advantage in at least one domain: in research. Late, very late, I discovered the true nature of science, of how it proceeds, of the men who do it. I came to understand that, contrary to what I had believed, the march of science does not consist in a series of inevitable conquests, or advance along the royal road of human reason, or result necessarily and inevitably from conclusive observations dictated by experiment and argumentation. I found in science a mode of playfulness and imagination, of obsessions and fixed ideas. To my surprise, those who achieved the unexpected and invented the possible were not simply men of learning and method. More than anything else, they possessed extraordinary minds, enjoyed the difficult, and often were creatures of amazing vision. Those in the front ranks displayed exotic blends of passion and indifference, of rigor and whimsy, of naïveté and the will to power, in a triumph of individuality.

Starting to work in André Lwoff's laboratory at the Pasteur Institute, I found myself in an unfamiliar universe of limitless imagination and endless criticism. The game was that of continually inventing a possible world, or a piece of a possible world, and then of comparing it with the real world. Doing experiments was to give free rein to every idea that crossed my mind. It was constantly to set up new "little lights." Every morning I ran to the laboratory to set up my experiments. It was a sort of bulimia; of delirium; like a kid unwrapping an unexpected toy. In the morning, I prepared the bacteria and the viruses I was working on, and ran the experiment in the afternoon. The next morning I got the results just in time to put together a further experiment to run later the same day, and so on. A fiendish pace. A race without end. The mad pursuit of the day after this one. What mattered more than the answers were the questions and how they were formulated; for in the best of cases, the answer led to new questions. It was a system for concocting expectation; a machine for making the future. For me, this world of questions and the provisional, this chase after an answer that was always put off to the next day, all that was euphoric. I lived in the future. I always waited for the result of tomorrow. I had turned my anxiety into my profession.

Night has swallowed up time. Serenely it shapes the silence out of which words emerge, and into which they return. And yet silence may occasionally split, without warning. There, where least expected, a bubble bursts at the surface of awareness, the bubble of a memory, a conceit, a desire, a humiliation coming up from the depths and reviving a vanished world. What a bag of tricks memory is! What a snare for images! What we seek in memory, we do not find. Rather we find there what we have not sought. Memory speaks endlessly of places, events, people—but of me, not a word. The thing to do is to proceed at your own pace, lingering here, fooling around there. Memory has little tolerance for being rushed. It is like those old men who possess some piece of information one absolutely must have, and whom one has to listen to for hours about their youth, or their love affairs, or their exploits as soldiers, in the hope of their eventually getting to the point. So, what is called for is patience. You must lie in wait for a sudden bolt to swoop you to your parents' home, to the bed of a former lover, to a desert foxhole under bombardment. You must let yourself float along in your past, the museum that no one else can visit. But don't harbor too many illusions: what comes back readily are the lessons gone over and over, the films shown again and again.

The return of the cold. The nights that do not end, the leafless trees shivering with boredom: all the signs of winter's onset grow more oppressive with each year. Each year, they proclaim more harshly the imminence of another cold, of another night, and summon again the long cortège of ghosts. In me the signs of winter set up an almost unbearable tension between what appears in my conscious mind and what is obscurely stirring within me.

I stole out of the laboratory. I left the Pasteur Institute by the back door and walked aimlessly through the streets. In the middle of the afternoon, dusk was already settling over Paris, drowning all in gray: stores, traffic, passersby. Men and women strolled by, unperturbed. A procession of people condemned to death, endlessly coming and going, whirling around and setting out again without hope of breaking free. I walked along a boulevard. I went through a park. I saw the Seine, dense and black behind the locked chests of the booksellers along the quay. I crossed a bridge in the stink of a traffic jam. I fled my reflection in a hairdresser's window. At the corner of a square, I was grazed by a cyclist who yelled back at me. I went round and round in the narrower streets. For a long time, I walked in this way without quite knowing where I was or where I was headed. Between low houses. Between huge apartment buildings.

And suddenly the avenue took on a direction. Abruptly I was on familiar ground. My step reverted to a familiar rhythm. Throughout my body, I felt the measured tread that had swung along this route many times. I was back on a well-traced path, a route I had taken thousands of times in one direction, thousands of times in the other. Running. Ambling. In good weather. In the rain. Alone. With friends. With pride after getting a good grade. With the fear of being questioned. Around me, everything settled naturally into the proper order, fitting in on the way from my parents' home to the lycée. The chestnut trees of the Place Malesherbes with, at its end, the building where we lived. And farther, the long reddish edifice of the Banque de France, which once seemed to me to house all the wealth of France; the rue de Phalsbourg with, in the distance, the rotunda of the Parc Monceau; the subway entrance and the bench where we endlessly discussed local or world events; the small street leading to what was then the National School of Business Administration; the rue

Cardinet, where one evening Roger D. and I found ourselves walking behind our history teacher who, accosted by a street-walker, brandished his umbrella and shouted, "Stand back! Stand back, lady of the evening!"; the two rival bookstores; and then, above all, the long somber building of the Lycée Carnot. In fifty years, all this had scarcely changed. The whole scene was still in keeping with my old album of internal images.

The door of the lycée was open. I went in. On the right, the same concierge's booth. In my time, he answered to the unusual name of Poléon and sold licorice in coils, which we would eat by the yard. On the left, in a small garden, the monument to the war dead. At the far end of the corridor, the narrow stairway. And finally the hall, the great place of assembly, the lycée's agora: a long, ill-lit, asphalt-paved rectangle whose upper two floors are lined with classrooms. At each corner, a stairway goes up to the gallery running the length of the second floor. At that time of day, the place was nearly deserted. Over there, three teachers walked back and forth, talking. Over here, little knots of students were going over each other's notes. The lights were on in only two or three classrooms. And the grimness of it! It was just as grim as I remembered.

What force, on this evening of solitude, had drawn me here? What was I seeking in this lycée I hadn't set foot in since the war? Some trace of the child or the teenager who had spent ten years of his life in this place? I looked for this trace, while walking around the entrance hall; while going out to the playground, up to the second-floor gallery, into the classrooms; while sitting after a fashion at one of the desks, now too small. Before me filed a parade of ghosts, students, but especially teachers. Monsieur V., who was stone deaf, but would yell "Quiet!" in the rare moment no one was acting up. The enormously tall Monsieur F., now nice guy, now fire breather, moving without warning from mildness to anger. Monsieur D., whose too-evident taste for young boys obliged us to pile up chairs in the aisles to prevent him from getting through. Monsieur G. and his Latin quizzes: "As many feet as hands. As many hands as feet. So many hands, so many feet. So many feet, so many hands. Turn in your papers.'' A weird dance of faces and tics. And, above all, what sadness in the gray walls and dirty bricks. What endless and unrelieved boredom! Ten years in this dump! What a place for a child to be sentenced to!

For more than an hour I tried to take up the motions of an earlier time. I walked to the right and to the left, from class to class, from teacher to teacher. But I could recapture only from the outside the lycée student I knew I had once been. I saw him as just one face among many in the class photos with the pupils grouped around the teacher, the smaller ones seated in front, the taller ones standing behind, and the names of each written by me below; and my own face surprises me today as much as the others do.

One impression came from afar, however, when my zeal to see everything led me to the lavatory by the playground. At last, there in the musty odor of disinfectant and urine—a definite memory. At this spot in the playground, during a recess, seven or eight of us eleven-year-olds are teaming up to play cops and robbers. Choosing the sides is G., a tall, fat boy with curly blond hair and eyes of forget-me-not blue. Suddenly he flings at me, ''A dirty Jew like you has to be a robber.'' Abruptly, the word *juif*—which, up to that moment, had had for me a rather high, thin sound—takes on the sharp, biting edge of a siren, a hissing that contains a judge, a slap, a needle. I redden, first from humiliation, then from rage. I have some misgivings about this fellow, so cocky and probably stronger than I. But if I give in, if I submit once, just one single time, now, he will never let me be. He will pick a fight with me every day. The other fellows watch us. And then I sense other gazes: my father, my grandfather, all the males in the family who so obligingly tell their war stories. So I jump on him. G. does not expect it. In his mythology, I can be only a coward, a chicken. It is one of the most savage fights I have ever been in. At the height of it, we find ourselves at the doorway to the lavatory. Bracing myself and summoning up every bit of strength left in me, I send G. rolling into one of the shit-holes, just as the bell rings to signal the end of recess. Stupefied, the others look at me with a mixture of disapproval and respect. Today, my pride in this triumph, not just over the other, but mainly over myself, over my fear; my pride in not having given in, has perhaps lost none of its force in fifty years.

Though in a flash I could recover the lycée student I had been, it was not in a series of images and events frozen in memory. It was in a series of emotions felt anew, relived almost as before. The rest, the lycée, the classes, the teachers' quirks, my friends' behavior, the immense boredom, even the fight with G., everything I know about these years, I could have learned from another source, such as an old schoolmate's reminiscences. But not the humiliation of the blackballed little Jew, not the struggle against my own fear, or the pride of success, of almost military victory. That no one could convey to me. That way, and that way only, I found not the memory but the very presence of the little boy I had been. That way, I had for a moment connected with my childhood.

As a child, I sometimes dreamed of being someone else. I even remember dreaming one night of myself as a hero killed in the Great War. The next day I had a rock-solid belief in the transmigration of souls. For a long time, while going to sleep, I feared I would wake up as someone else. To combat this fear, I settled flat on the bed, my cheek well centered on the pillow, the sheet pulled taut, my body motionless. Everything had to be impeccable. It seemed to me the order of my bed guaranteed the order of the world. I struggled against falling asleep so as not to forget myself. As in going over a lesson, I passed in review everything that seemed to characterize and define me: my mother, my father, my grandparents, my room, my things, my books, my class, my friends. . . . The more items I added to the list, the more exact the self-description seemed. But soon this parade acted on me like the sheep one counts to go to sleep. Everything got muddled. I lost consciousness. When I awoke ten hours later, I seemed to have slept only an instant. So I thought it perfectly normal to find myself the same, for I had scarcely had time to lose myself by forgetting. Though the fear of becoming someone else has disappeared, I am still sometimes surprised to wake up and find myself the same. The persona that has come undone in sleep reconstructs itself on awakening. How is it that this busy stir always re-creates the same individual, the one I call myself? How is it that, among the thousand personalities that one believes it possible to be, it is always the same one that prevails, and not another now and then? Why doesn't the system slip so that, after sleep has disassembled the mind, its memory and will, the mechanism is not reassembled somewhat differently, to form a different person, a different me?

What intrigues me in my life is: How did I come to be what I am? How did this person develop, this *I* whom I rediscover each morning and to whom I must accommodate myself to the end? But how am I to encompass this life? How even to describe it? By a sequence of events, a series of situations, of anecdotes fitted into a chronology? Yet memory does not reconstitute life and its unfolding. It casts will-o'-the-wisps over the enchantment of a brief holiday while shrouding in a cloak of sadness the tranquillity of sustained work. With an invisible projector, memory lights up unknown continents and brings into view unsuspected lands. Certain memories reflect flashes of imagination; others, flashes from the past. A past that I must patiently solicit. A past through which I must each moment pick my way among the memories that court me and those that run away.

If, then, in this series of images my memory presents, I rediscover my life imperfectly, how am I to define it? By some continuity or permanence? Surely not. Seen from the outside, the life of someone harboring a unique passion may seem as straightforward and purposeful as the trajectory of an arrow. For a long time, I admired the heroes who advanced unswervingly along a single path, whose existence could be summed up in a formula, who never stopped fighting *for*—for the grandeur of their country, for the glory of their God, for freedom, for independence, for the progress of science, for the conquest of power, for the accumulation of wealth. Or those heroes who never stopped fighting *against*—against the neighboring country, against the God of other people, against poverty, against ignorance, against privilege, against inequality, against a tyrant, against evil. With age, however, I learned to distrust these heroes, their perfection, their absolutes: at first because, with Freud's reconsideration of Homer and Plutarch, no one past puberty can be unaware that behind the hero there usually lurks a monster; but also because history shows all too well that nothing is as dangerous, as murderous, as ideology, as fanaticism, as the certainty of being right. Nothing causes as much destruction, misery, and death as obsession with a truth believed absolute. Every crime in history is the product of some fanaticism. Every massacre is performed in the name of virtue; in the name of legitimate nationalism, a true religion, a just ideology, the fight against Satan. I love fixed ideas, but only if they can be changed.

I see my life less as continuity than as a series of different selves—I might almost say, strangers. At the end of the line, I see the little boy, the only child, cajoled by a sweet mother, spoiled by all, playing (too often alone) while mouthing words that he tries out and twists ad infinitum. Then comes the adolescent, swollen with vanity, full of ambition, a shade backward with girls. Then, the medical student preparing to become a surgeon. The fighting man of the Free French forces, thrown across North Africa with an infantry battalion from Chad. The wreck of a man, lacerated by grenades, who returns to Paris and makes a stab at ten different professions. The beginner at the Pasteur Institute, discovering in awe the world of research and biology. All this gang marching in single file. I have trouble imagining that, when the name François Jacob is called, all these selves can leap up and answer, "Present." When I come across my name in a report card, a military document, or an old newspaper article, it seems to refer to fellows who happen to have the same name as I. Would I recognize them if I passed them in the street? Like the bird contemplating the shell it has just broken out of, and saying, "Me? In there? Never!"

Recovering each of my past selves takes time and effort. I must concentrate, prepare myself, rather like setting out to visit old, long-neglected friends. Or, rather, like planning a trip in foreign countries one has not returned to for years. Upon arriving in each country, one has to relearn the geography, the customs, the manners. Each time, one has to devote oneself to the unrelenting observation of existence, in a different texture of space and time. Each time, one must relinquish the habits of the last country before acquiring the habits of the new. Each time one must relearn a particular milieu, particular faces, a particular language. Similarly, for each of my former selves, I must, by paying close attention, become accustomed to seeing again the things, the people, the future of former days. To finding again for each of them the little vanities, the fears, the reasons for pride or delusion, the familiar echo. And the contradictions. And the idle wishes. And the music of the mornings that launches one into the day. And the dreams rising at night to fasten on a face, a toy, a body. And the way of going to bed, of getting up, of talking. From the unending incantations of the little boy playing with words, I must move on to the prideful jargon of the medical student who couldn't clear his throat without tossing off a technical term. Or even go back in time from the young researcher wanting to emulate Pasteur, to the fighting man who swore only by de Gaulle. Then to the little boy who took himself now for Napoleon, now for Achilles, son of Thetis and king of the Myrmidons. To find the rhythms of particular times, I must listen for the bell that marked the flow of classes and recesses. Or wait for the ring of the phone that woke me during my nights on call at the hospital. Or listen for the interval between the departure and the arrival of the shrapnel shells. Or evoke the vibration of the old blender where the bacteria were mating. One has to retrieve the nights that ended too soon, the dawns that did not come soon enough. But how to reconstruct worlds from which has disappeared the uncertainty of the future, a future now in the past?

What interests me most about these personas of my life are their transformations and interrelations. How did the young man emerge from the adolescent, and how did he replace a carnivore's vision of the world with an intransigent idealism? By what twists did the adult end up changing values, submitting to the general interest, and playing a game that a short while back he was vehemently denouncing? Also intriguing to me are the relations between the current *I* and these former selves. In what way is the professor at the Collège de France still the medical student? Can he still choose to be that student? And does he want to? Am I still one with the fighting man of twenty, or have I betrayed what he hoped for or scorned? Can it be that I still believe in, or even recall, what gave sense to my youth? And would the adolescent be in the least concerned with what matters to me now?

And yet, as different as these selves making up my life may seem now, for more than sixty years they have, every morning upon awakening, recognized each other. They have found each other the same, remembering the same name, the same past, the same condition, the same face, the same troubles. Each time they have had the feeling of resuming the same role, after the nightly intermission, at the exact point it left off the night before. This consciousness of unity is not only that of my body, its habits, its inclinations. Even more, it is made of those memories that travel through time in flashes, that whisk me into the past, accurately drop me into a situation experienced decades earlier, generate in me a sudden intense sensation, like a sign that I alone can detect and decipher. Such is the sensation caused by the image of Cleopatra or of the Greek goddesses; or again, the deep humiliation I felt on 17 June 1940 in the little inn in Auvergne, when the words of Marshal Pétain gave definite notice of the fall of France and acknowledged the failure of everything that at twenty I had learned to believe in; the whole idealized construct of a country, a state, an army, which had until then seemed as solid as the world itself and, in a month, fell to dust. Other memories of equally violent sensations, other humiliations, other rages, and also pleasures, remain buried in the deepest part of me, ready to burst forth when a thorn accurately pricks the vulnerable spot. This close connection between my memory and my sensitivity, between certain recollections and my tendency to react to them powerfully now, unites me unequivocally with my past selves, since the same provocations occurring today would not get a rise out of me. All those old humiliations, shames, and rages even make me smile, now that the harshness of emotion mixed with memory has faded. Isn't this how I have learned certain aspects of life that cannot be taught? Don't we have to experience certain unpleasant, even hateful or ridiculous, situations, and endure confusion or anger, in order to impress on ourselves what no word can teach? Is there a better instructor than self-love on the path that no one can take for you?

This communication between my heart and my memory, all these emotions well up when awakened by some recollections of the past. They weave a web that links what I am with what I was. They compel me to unity, as does the feeling of an old complicity with all my past selves in those eternal debates that make up the inner dialogue. Or the certainty that each in turn gave me a cue. That each played a role in the struggle of angel with beast; in the pas de deux that directs one now toward things in desire and now away from them in disgust; in the endless combat between what in a man stops him, freezes him, fixes him, devotes him to the particular, to a woman, a destiny, a premature old age, and what, on the contrary, renews him, changes him, makes him leave, propels him toward a hundred existences, a thousand facets, all women, the universal, eternal youth. The conflict of his tendencies, his practices, and his name. If I have forgotten, it is because the same is no longer the same.

And then, how not to see that all these selves of my past life have played the greatest role, and the greater the earlier they came, in the development of the secret image that from the deepest part of me guides my tastes, desires, decisions. Starting in the younger years, imagination seizes on the people and things it encounters. It grinds them down, transforms them, abstracts a feature or a sign with which to shape our ideal representation of the world. A schema that becomes our system of reference, our code to decipher oncoming reality. Thus, I carry within a kind of inner statue, a statue sculpted since childhood, that gives my life a continuity and is the most intimate part of me, the hardest kernel of my character. I have been shaping this statue all my life. I have been constantly retouching, polishing, refining it. Here, the chisel and the gouge are made of encounters and inter- actions; of discordant rhythms; of stray pages from one chapter that slip into another in the almanac of the emotions; terrors induced by what is all sweetness; a need for infinity erupting in bursts of music; a delight surging up at the sight of a stern gaze; an exaltation born from an association of words; all the sensa- tions and constraints, marks left by some people and by others, by the reality of life and by the dream. Thus, I harbor not just one ideal person with whom I continually compare myself. I carry a whole train of moral figures, with utterly contradictory qualities, who in my imagination are always ready to act as my fellow players in situations and dialogues imprinted in my head since childhood or adolescence. For every role in this repertory of the possible, for all the activities that surround me and involve me directly, I thus hold actors ready to respond to cues in comedies and tragedies inscribed in me long ago. Not a gesture, not a word, but has been imposed by the statue within.

II

THIS GIRL OF TWENTY, smiling gravely in her nurse's uniform, is my mother. The photo was taken in a First World War military hospital, some months before the return of peace, some months before her marriage. The person who will be my mother is standing there, all sweetness, all innocence under her white cap, rolling a bandage. The tenderness in her face. The texture of her cheeks. The warmth one would feel lightly brushing them with one's lips. The play of subtle curves in those fingers that one guesses are sure of themselves, nimble, patient, skilled. But, above all, the searching look whose anxiety betrays the serenity of the face; the truth of those eyes which every day looked on suffering, misery, death.

For this army officer's daughter, raised in the protected, cozy world of the provincial *petite bourgeoisie,* was abruptly sent into the war hospitals, into the halls of torture, anguish, and obscenity that were the wards of the wounded. How did she bear the howls of pain? And the fear of death? And the smell of putrefaction, of ether and gangrene. What were her hopes and dreams, faced with the bloody, mutilated bodies, her first young men's bodies, delivered naked and gasping into the hands of surgeons? Was she in love with a hero? Or with an anonymous heroism? In this world of bedlam and resignation, it was not enough to get on with washing bodies, changing bandages, handing out medications; you also had to keep up the struggle against renunciation and despair. Above all, those boys of twenty had to be taught to wait: to wait for revival, for the night to end, for the doctor, for medicine, for a visit, for a remission, for a smile. They had to be rescued from the nightmare, accustomed to live again, almost reborn. Barely out of childhood, the little nurse found herself in hell. As she confronted mutilated limbs, shattered faces, vacant eye sockets, what meaning did the words *sacrifice, homeland, valor* have for her? And how did she react to the coarse language, the casual intermingling, the desire of these men? Did she sit by their beds to listen and pour out a bit of dream? Did she have any romances, and how far did she let them go? What did she believe in? What were her hopes? What passion stirred her, giving her expression in the photo this nearly unbearable mixture of distress and serenity?

My mother's charm, her simplicity and beauty attracted much tribute from men. Some years ago, I met an old suitor of hers, a man of eighty, who had been a high civil servant and one of the architects of our recent republics. My mother's contemporary, he spoke of her with a warmth and an emotion that could come only from the memory of an unfading past. When I met him, his feeling for my mother must have still been strong enough to move him, and to move me, in speaking of her in this way.

The maman of my childhood: Tenderness. Sweetness. Perfumed warmth. Safe harbor from all fears, against all violence. Maman who taught me everything: To look at the world and to speak. To stand up on my hind legs and walk without hands. To dress myself and tie my shoelaces. To brush my teeth and knot my tie. Maman, who rocked me to sleep, bathed me, wiped me, blew my nose, disciplined me, tucked me in, caressed me, scolded me, watched over me. Who took me to school, where at first I didn't want to see her leave, later didn't want to see her stay, and eventually didn't want to see her come. Maman, who for years launched me into the day each morning. Who waited at her window to welcome me home from school each afternoon. Who, when I was a medical student and would get back late at night, always left a snack on the table with a note as tender as a kiss.

When I was ill, maman rarely left my side, her forehead creased in a frown until my uncle Henri, immediately alerted, arrived. Uncle Henri—my grandmother's brother, the same one she confused me with on my final visit—was a fine figure of a man, one of the central personalities in the family, a kind of thunderous Jupiter, a giant with a long grizzled beard, and huge ears, like my grandmother's, but mainly with great good nature, great warmth in the crinkle of his eyes and the sparkle of his gaze behind his glasses. This doctor, who had only a nodding acquaintance with medicine and relied on common sense, infused his patients with his optimism and high spirits. He loved good food and women. It was even suspected that he took a close interest in his prettier patients.

After joking with me, Uncle Henri would put a cloth on my chest. And when he pressed his ear to the cloth to listen to my breathing, his beard tickled me. During the auscultation, maman walked on tiptoe, trying not to make a sound. Then came the prescription ceremony. Afterward my mother, a bit anxious, tremulous as a girl hoping for a compliment from her dancing partner, saw Uncle Henri out, seeking assurance that her little boy had nothing serious. She then hurried to the druggist to fetch the miracle drugs that were going to set her offspring back on his feet. And only after measuring out the prescribed dose, after watching closely over my compliance as I swallowed it, did she allow herself to relax and play with me.

Sweetness of a vanished world. My childhood room, with the wicker table and chair, the toy box, the plush elephant, the lead soldiers, the smile of the little old lady across the street who waved to me from her window. Joys of long convalescences that kept maman nearby. Smell of eucalyptus. Spoonfuls of medicinal syrup. Thermometer. Night light. Herbal tea. Decalcomanias and seven-league boots. All the symbols of my childhood, birthday cakes and candles, the illustrated papers on Thursdays, hoops, merry-go-round, and seesaw. Dirty hands. Skinned knees. Scratched scabs yanked off too soon. Start of a new year in school, with new books and notebooks that maman covered with colored paper. Smell of my leather school satchel. Agate marbles and steel-nibbed pens. Afternoon snack. Afternoon at the circus with its smells of stable and citronella. Fables of La Fontaine recited, droning, stumbling. Pages of handwriting. Waiting for summer vacation. Waiting to grow up. Waiting for maman to return, for the sound of her step on the stairs, for her key in the lock.

It has been a long time now since she last returned. Since she last spoke, she who was so good at telling stories, sitting on my bed at night before tucking me in, always mixing stories with words of encouragement, spreading armfuls of confidence and security. All during the first year of the Second World War, she struggled against illness, gradually weakening, each day walking with greater difficulty. And our roles were reversed. It was I, the young medical student, who came to sit on her bed, telling her stories, discussing diagnosis and treatment, awkwardly trying to reassure her. In a final proof of her love, she pretended to believe me, to instill me with confidence one more time, one last time. She quietly died in early June 1940, at the height of the storm. They took her away. They lowered her in a hole. And I still sometimes hear the first spadeful of earth land on her coffin.

She died in time to know nothing of the horror, to elude the invasion of the barbarians, the shame in the street, the running away from the yellow star. She died at the very moment a world was collapsing; when, in order to fight, I had to get out of France. In my memory, France's defeat and my mother's death have merged into a single event. There, my childhood came to an end. Afterward, everything went on, but nothing was the same. All during the war, I forced myself to forget my childhood so as not to think about maman. I could not let her become too alive, too beautiful, too terrible. I had to keep her from remaining a constant presence, a too familiar ghost, a fever smoldering in the pit of my sleep. I had to keep from thinking that she was still alive, that she was waiting for me, concealed from the Gestapo. That my return to France would bring me back to her. In the end, I succeeded. My mother's features, her voice, her silhouette grew blurred. I buried her memory at the bottom of an ocean where everything changes: rhythms, densities, colors. I drowned her image in the deeps of my tenderness.

Evenings and mornings of my childhood. Evenings of fear after maman's kiss. Once the door was shut, the darkness and the creaking in the wall. The tall woman in white draperies across the street who skipped back and forth on the rail of the balcony, waving her arms. The sounds of beasts scuffling around my bed. The steps of robbers out in the hallway. The silhouette of a witch behind the door. My head beneath the covers so as to see nothing, to hear nothing more.

And immediately afterward, the next morning: for a night of sleep seemed to last but a few minutes. Waking up was like surfacing after a plunge in the soft dark world of the sea. The departure from a country peopled with strange inhabitants. A country where time changes course and what is possible changes register. Where things easily take on the look of animals and men of women. Where desire detaches itself from memory. The return to the confines of life. A still vague murmur. A confused hum gradually becoming audible from the awakening street. The diffuse glimmer of daylight. Everything announcing the return to the dry, hard world of the real gave me, like an animal, a quiver in every fiber of my body, a kind of primitive feeling of existence.

At that precise moment, before resuming my identity, while my mind was free from any past, at just that instant, the game was beginning. I had to do nothing to rush things, to hasten the awakening. I had to remain motionless, flat on my stomach, my arms clasping the pillow, eyes tight shut. It was forbidden to move a muscle. Before I could perceive the world around me, see it, hear it, I had to reinvent it and set it in place, as much by imagination as by memory. In immobility, in the stillness of the morning, I started mentally to do some heavy moving of furniture, walls, houses, streets, cities, continents. To begin with, I reconstructed my room: it organized itself around my bed, a sort of impregnable citadel, a refuge from violence and the nastiness of the world. The window had to be placed first. I would try putting it on my right, then on my left. But the only place it fitted was over there, toward my feet. As for the door to the hallway, it whirled along the walls before coming to stop near my head, a bit to the right, almost within reach. Before putting the ceiling on the box I had constructed, I might arrange the furniture in it, trying the most varied combinations, as in a mock-up of an apartment one is about to move into. What, all things considered, fixed for me the positions of table, chairs, and armoire was my body's knowledge of the surrounding space. It was the amount of effort I had to expend to reach one or another of these objects. As soon as my room was reconstructed, I rediscovered my territory. It was my cave, my theater, the field of maneuvers endlessly traversed every day in every direction.

When, eyes still closed, I had managed to construct this room from nothing, when it existed in my mind down to the last detail, I had to arrange the rest of the apartment around it. On the other side of the wall by my bed, the bathroom came to settle. Beyond was the kitchen. What enabled me to position my parents' room was at first a confused murmur, out of which gradually emerged voices among which I began to distinguish that of my mother and that of my father. Thus reconstituted, the apartment came then to fit on the fourth floor of the building to which the first hiccups of the elevator began to give life. Once the building was constructed and securely set on its foundation, I could border it with a street whose first sounds I heard coming up through the window over there near my feet. It was the quality of these noises—how the cars' tires, the horses' feet, the heels of passersby resounded in the dry air or were smothered on the wet ground—that showed me the way to color the street. To illuminate it with the first beams of a triumphal sun. Or to drown it in fog and mist. When I had thus situated my room in the apartment, the apartment in the building, and the building in the street, when I had decided on what kind of sky to cap the landscape, and on what color to paint it, I could complete the final stage in my construction: the placement of the Seine, which I oriented in relation to the street we lived on; and which, after some groping, allowed me to arrange around myself first Paris, then France, then Europe, then the earth, the sun and the moon, the stars. In short, in a few minutes I would re-create the universe.

Bestirring itself this way, making matter and objects swirl around me, my mind had finally become completely awake. My mind but not my body, which I kept carefully numbed. Up to then, indeed, I had remained perfectly still, eyes closed, muscles relaxed. Only after the world was reinstated around me in my mind did I resign myself to opening an eye to check the arrangement of my room, the position of the table, the toys, my things. Only then did I grant myself the right and the power to move. First, by a sort of general quivering in which I hugged the pillow and snuggled into the mattress, enjoying my bed for one last moment. Then I focused my attention on a single limb: my right arm, say. I began cautiously to extend it. To feel each muscle, each joint, each inch of skin. To bring into play, one after the other, my hand, my elbow, my shoulder. Mastery of this arm reassured, I went on to another limb, then to a third. But the last one, let's say the left leg, I let loll for a while. I prolonged its immobility while directing to it all my thoughts. This limb then became so dense, so cumbersome, it seemed weighed down by a cover so heavy that I thought I would never be able to move it. It was only by a huge effort of will, a tremendous outlay of energy, that I finally managed to rid my leg of its inertia. Then at last I had the right to get out of bed.

Shouts and anger. Fist on the table and beetling eyebrows. My father's flush and my mother's pallor. Throat constricted, heart heavy, I can't get my cold mashed potatoes down. I shrink down in my little chair, softly weeping. My father and mother are quarreling. Or rather my father is scolding my mother. He is shouting. She overbuttered the cutlets. She knows butter is bad for him. She doesn't give a damn about his health. She does just what she pleases. She doesn't look after him. Maman doesn't answer. Her face set, her mouth tense, she looks off in the distance, with an occasional wan smile in my direction, as if to excuse herself, as if to excuse him. I am scared. Scared of him. Scared for her. Scared of catastrophe. Scared of chaos. I hate him. Oh! how I hate him. We would be so good the two of us, maman and I. I shrink further in my chair, sniffing spasmodically, but making no sound so as not to attract attention. I long for an end to this argument. For Papa to stop shouting. For the return of quiet and harmony. For everyone to love each other. I rock back and forth on my chair, slowly at first, then a little faster. Suddenly I rear back. Screaming, I fall backward, taking care to avoid getting hurt. The shouting stops dead. Mother and father dash to help me up, to placate me, to make sure their little boy hasn't come to any harm. In tears, I bury myself in maman's arms, my nose in the hollow of her armpit. The fear they both had for my safety reconciles them. They kiss. I am quite proud of myself. I have defended maman. I have saved her from violence and injustice. I have brought peace.

Silence when I come near. Little smiles. Winks. Snatches of conversation fluttering up there over my head. What I am not supposed to hear. Inane and irrelevant answers to some of my inquisitive questions. A feeling of being excluded, of being kept away from a secret I didn't have the right to share in. "Later on, when you're a grown-up." But a child's ear has an acuity, his mind a shrewdness that adults have long forgotten. As they have forgotten the child's anxiety when struck by a blatant discrepancy between words and actions, between morals and behavior, between the advice of others and personal experience. Forgotten the confusion when a gap appears between what people say around the table and the secret life, the life of the glands, the senses, the viscera. When, from behind the smile, the self-assurance, the strength of adults, begin to well up cares, failures, miseries. The discovery that there is no father without some weaknesses or, later, no society without prisons. The transition from myth to the human condition, the terrible moments of truth in the child's constant effort to construct his world.

My earliest images of my father are of a valiant knight, a hero of legend. We are walking in the street, my hand in his. He shields me from every danger: from the Indians lying in wait behind that porte-cochère, from the enemy cannons stationed around the corner, from the wild beasts that roar when they see me. Or, we are both on horseback: he is holding me in front of him on the saddle, and, when the horse trots in parade at the head of a line of troops, I am petrified, but wouldn't give up my place for an empire. Or again, at the fair, he has me shoot a gun at the enemy soldiers waiting for us at the back of the booth. He was, as it were, my founding image, my model of the human being on this earth. In him were allied all the virtues: boldness, strength, courage. And elegance, with fine limbs, long hands, long sallow face, slender nose, small mustache and jet-black hair, dark eyes. The conquering look of the artillery sergeant in a fancy képi who can still be seen in a blue pastel drawing made one day at the front by a regimental comrade, who was killed the very next day.

With his two brothers, my father was a partner in the small real-estate firm founded by my grandfather. He went each week to Normandy, disappearing Monday morning and returning Tuesday or Wednesday evening. Maman was always a little sad in his absence. I, however, was proud of having a father who traveled. The world stretched beyond the house, beyond the city, beyond the nation, along a mysterious universe which he explored. Mother at home, father bestriding the world: this appeared to be the order of things. I felt for my father an admiration, a veneration mingled with a fear, which increased when, to reward me, he allowed me to visit his office. For me this was a sort of inaccessible shrine. I entered it as one enters a temple. Its solemnity was reinforced by the thick rug which muffled the sound of footsteps, by the majestic armchairs covered in brown leather, and, in the middle of the room, by the great table spread with books, notebooks, documents. My father settled me in a great armchair and talked about his youth, his early friends, the Dreyfus affair. He told about how, at the lycée in Nancy, he had struck one of his schoolmates who had called him a dirty Jew. How he'd gotten away, thumbing his nose at the classmate's father, a cavalry officer with a riding crop, who was waiting for my father at the exit. He told me war stories: the artillery, the horses, the nights spent in the mud, the hunger, exhaustion under bombardment. How could I resist all these exploits? I drooled with admiration. And besides, he knew everything. This magician could do anything: hold a cane balanced on his chin; make a coin vanish and reappear on his nose; spin his wedding ring like a top; build great castles of cards which I blew down in a burst of laughter; imitate Charlie Chaplin's walk; fly paper airplanes; make a toy come out of his hat. There was nothing he couldn't do.

My father often took my mother and me to antique shops and auctions. He loved old things, bibelots, paintings. His fairly modest means obliged him to limit his acquisitions. But I can still recall objects in my childhood home he had picked up here and there: a large black and white vase, an old pendulum clock with metal leafwork and porcelain figurines, a bronze lion playing with a ball, a Japanese dagger I was forbidden to touch. I see drawings: animals, ships, farm folk at work. Also English engravings with horses and hunting scenes. A few paintings. Primarily two that were in our home as far back as I can remember. First a nude, one of my earliest images of woman, seen from the rear, standing, the head inclined on her bent left arm, leaning against a rock. A fine nude: very smooth, very pink, very fleshy;

its plenitude well displayed, as were its curves; vibrant in the masses of its muscles, in all its shining roundnesses; an opulent body, sensual even in each lock of hair, in the mother-of-pearl ears, in the contour of the shoulders; a body that sang of all its transparent cleanness; a body as naturally made for nakedness as a flower. This scene was said to represent Andromeda waiting on her rock for the arrival of the monster. What troubled me about the picture was not just the glory of this body, which had upon it the dew of dawn; but rather that it was captive and destined for the pleasure of the monster. It was the contrast between the luster of that flesh and the horror of its fate, between the triumph of that nudity and the torture that awaited it. For me nothing seemed able to surpass this painting in which beauty and sensuality found themselves closely allied to cruelty and mortification. Until the day my grandfather, the general, pointed out to me a number of imperfections: a thigh too long, a misshapen hand, shadows that bore no relation to one another. But at the same time he showed that each of these aberrations had a function: the anomaly of the hand conveyed the woman's powerlessness; the exaggeration of the left thigh going behind the right slightly unbalanced the body and emphasized its constraints. As for the contradictory shadows, they tended to disorient the viewer and to give him a feeling of anxiety.

The explanation continued with the other picture I remember from my parents' home: a landscape near Paris, with the Seine in the foreground flowing past gardens and orchards. In this I was entranced by the delicacy of the colors, the blurring of the contours which gave an effect of mist over the river, and particularly that delightful discovery of the Impressionists: splashes of light dancing on a lawn. Since my childhood, this canvas has irresistibly evoked for me the heat of summer, the buzzing of flies, the fragrance of honeysuckle and geraniums in the garden of our summer place in Etretat. But, there again, my grandfather pointed to the liberties the painter had taken with reality. In several places, for example, he had drowned all outlines in a haze of water and sun that suppressed all demarcation between the river and the riverbank. This trick gave the viewer an impression of the blinding sun, of light ricocheting in one's eyes. What my grandfather thus showed me was that reality is not given. That the artist makes his account of it. That the world does not come ready-made. That it is up to each person to reconstruct it.

All my friends had brothers and sisters. Not I. I disliked this state of affairs. I felt it to be a wound, an infirmity. I saw it as a violation of the rules of the game, an injustice depriving me of an ally to which I was entitled, an accomplice in games and study, an associate for life. I didn't understand why my parents failed to remedy this situation. I begged them to give me a little brother and sister. But to my pleas they answered only with words of evasion. *Later on. We'll see. Maybe. If you're good.* I would have been glad to take care of this matter myself, but I did not see how to go about getting a baby on my own. The recipes mentioned in my presence seemed frivolous. Neither my friends nor I had ever seen a baby in a rose or a cabbage. The idea of a baby store also seemed unlikely, given the nature of the goods. So I saw only two possible solutions: either every baby was turned out by God Himself who, at the parents' request, sent it by a special emissary to whom a child had no access; or the baby was fabricated on earth by a sort of "Adam's rib" process—that is, by somehow taking a piece of the parents and fashioning it into a child. I leaned toward the latter situation, for it appeared to explain why there had to be parents to produce babies. An opinion strongly reinforced when I learned that the transformation of the piece of the parents into a baby happened to take place in the mother's stomach, a process that gave her a majestic shape. I needed a good deal of time, however, to understand how this fragment was obtained from the parents. Some friends of mine claimed that the operation occurred during a kiss. This seemed a bit far-fetched, for to cut off a sample of one or the other, even of both, implied some sort of bite in the course of the kiss. Further- more, it had to be admitted that the fragments of the parents so liberated were then swallowed by the mother in whose belly they were to be implanted in order to be transformed into the baby. Despite my attentive observation of the kisses I had the chance to witness, I never noticed any biting violent and bloody enough to persuade me that the kiss played any role in the making of a baby. Until I finally established some connection between the birth of newborn animals and the frolics of the dogs in the Paris streets, of the fish in my aquarium, of the chickens in the henhouse when we were on vacation. But, for a time, the oddity of the thing, the difficulty of imagining it in the human species, prevented me from giving it the importance it warranted, and from connecting it with certain of my favorite stories, with Aphrodite and her son Eros.

Being alone, lacking brother and sister, meant that often, too often, it was for myself and myself alone that I showed off—dancing in front of a mirror miming strength, intelligence, and grace; seeking to please, to tame, to conquer; acting the person who knows, commands, foresees; asking my ordnance officer to fetch marshals Murat and Ney to prepare for the next day's battle; winning the hundred-meter dash in the Olympics; declaiming the verses I'd just written; rescuing Joan of Arc from the stake; acknowledging that, thanks to me, peace and justice reigned on earth. In short, I strutted about like a peacock in full display.

Being alone, lacking brother and sister, meant that often, too often, I found no echo in house or garden; that I had neither prey nor predator to confront; that I encountered no opposition, no live resistance, no *other* able to represent the rest of the world, to be both my obstacle and my double, both rival and acolyte, both competing and familiar mind, both intimate and indispensable hostility. In short, there was no one to return the ball to me.

Being alone, lacking brother and sister, meant that often, too often, it was to myself and myself alone that I talked; that the one voice I heard was my own; that I repeated words ad infinitum, polishing them, chewing them, distorting them little by little; that I sucked on them like hard candy; that I ate them like little cakes; that I repeated them to exhaustion, to the point of nausea.

Etretat. A hot summer day. Our vacation home, near school, on the road to Criquetot. A small white house with wide wooden balconies painted green. The lawn with beds of geraniums. Thickets of black and red currant bushes. My parents have friends to visit, another couple. The men are taking a walk down the road. The ladies stay in the dining room to chat over an orangeade, happy to be together, happy to tell stories. Smiles. Low voices. Muffled laughter. I want to laugh too. While playing, I try to listen to what the women are saying. But they are speaking too softly, too quickly, for me to hear. I can catch only a few words, syllables: "playing tricks," "husband too nice," "utter madwoman," "some ambulistic." Ambulistic! Never heard this word. "Maman, what does that mean, ambulistic?" "Children should not listen to what adults say. Go and play in the garden, honey." I go to the garden, nibbling that strange word. Ambulistic. Some ambulistic. More ambulistic, less ambulistic. Not ambulistic. Are you ambulistic? Am I ambulistic? I am going to ambulist you. To ambulist a man. To ambulist a woman. Ambul. Istambul. Istambull. Horribull. Terribull. More bull. Ambull. Some ambul. Some ambulist. Some ambulistic. Some nambulistic. Somnambulistic. In the warmth of the late afternoon, I walk. I go all around the lawn, punctuating each syllable with a kick of a heel. The words I don't know I repeat, feeling and deboning them, reshuffling their elements until I know them by heart, as though by their mediation I would gain some access to the unknown, as though they would provide the materials I needed to control the world around me.

And then there were expressions, whole sentences, verses of songs that ran through my brain: "La Marseillaise," for example. When on public holidays people stood up and sang, *Marchons! Marchons! Kunsanguempur à brevenossion* [*Qu'un sang impur abreuve nos sillons*]," I felt myself gripped by some exotic music. Curiously, these incomprehensible expressions evoked for me images of the Orient: towered temples, bearded men in turbans, galleys with sails unfurled. On the other hand, when our record of *Carmen* got to "*l'Amour est enfant de Bohème*," it was always with some apprehension that I waited for the passage *"si tu ne m'aimes pas, je t'aime. Et si je t'aime, prangare datoit [prends garde à toi]."* I couldn't resist these meaningless words passionately sung by a woman's voice that rolled its *r*'s. I gave in to them directly. I let myself be invaded by the mystery, the magic of these words which, with a sudden fury, unleashed—from I know not what lair of monstrous images—nightmare visions in which hordes of enraged wild animals hurled themselves on bands of angry barbarians.

Power of words. Sweetness and violence of the words with which I tried to apprehend, to dominate, the disturbing world around me. By constantly acquiring words, by repeating them over and over, I seemed to grow more quickly. I felt that I was thus hastening the day when I would cease to be a child and would, finally, become a full-blown actor in that great spectacle that is the life of the world. To repeat words, to play with them this way, I felt as if I were in some way rehearsing the role that awaited me. A role whose nature I had scarcely the means to discern, but that I had no doubt would some day be of great importance.

I am seated on a train with my mother and father. We are traveling to Dijon to spend the Christmas holidays with my grandparents. I have a place next to the window. The train is speeding through a landscape dusted with snow. Forsaking my book and colored pencils, I amuse myself watching houses, gardens, telegraph poles, cars at the level crossing, farmers and their dogs rush toward me at full speed and instantly vanish. A strong impression comes over me: the conviction that this whole landscape, with its forests, fields, cows, has no continuity, no permanence; the feeling that in my absence it will not be there; that it was set up as the train approached, and will be taken down once it has passed. The setting up is done with precision, with an unremitting fidelity, since from one year to the next I find everything in place: the same hills, the same little stations, the same villages hugging the bank of the river. How is it that there is never a slip? How is it that the arrival at Laroche-Migenne never changes, even imperceptibly? That the tunnels are always the same length? And have the far-off and ominous hills of Alésia, whose approach my father has me on the lookout for on every trip, always been set in the same place since the time of Vercingetorix? Where was all this stored in our absence? If the train went faster, would it go too fast for the stagehands in charge of the countryside? Would it arrive in still empty places, surprising the nothingness that must prevail there? Even today, I am not sure that I do not hold traces of this idea of a stage set put up for our arrival, and dismantled when we leave.

All these movements of the landscape, these sets erected and taken down in haste, all this accorded with a strange and tormenting idea I had as a child. For a long time, an ill-defined belief in a sort of secret reality dwelled in my mind. There was the dream. There was the reality of every day. But, besides the officially real, the everyday, the recognized, there existed another world that, so to speak, duplicated this one but remained in the dark, clandestine, unofficial. It had nothing to do with stories of witches or fairy tales; nor with the hereafter of the Odyssey or the Bible. Rather, it was an aspect of the world that escaped observation by the usual means. There were found, waiting their turn to enter the maternal belly, future human beings, babies to be born. It was a sort of antechamber of the world, a space set apart to serve as understudy of the real, to give depth to the living. Perhaps this invention answered some need to give substance to life, a need that religions allow to none but themselves the right to satisfy.

It was mainly Claire who put these bizarre ideas into my head. She often came to baby-sit the evenings my parents went out. She was a tall thin woman with short hair, dark eyes, pockmarked skin, a supple bearing, skillful with her hands, reserved in her speech in the presence of my parents, waiting only for their departure to treat me, to my great satisfaction, to the most lugubrious stories. Before leaving, maman would get Claire and her knitting settled in the room next to mine. But no sooner had my parents gone out the door than I, instead of getting into bed, would race in to Claire to listen, late into the night, to her dim-witted horror stories. How could my mother—so anxious, so preoccupied with everything that affected me, so solicitous as to become quickly annoying—how could she, in this matter, let her attention wander?

I did not understand everything Claire said, but I could listen to her for hours on end. I could listen forever to that feverish, rough voice, whose harshness gave the stories a depth, a strength that impressed me. She talked very fast. Her stories all took place in her native village in the Pyrenees, a village cursed ever since the evil that befell its chatelaine. Most of the land belonged to one family whose sole offspring was a young count. He had married the loveliest, the blondest, the noblest, the gentlest, certainly the most wonderful of women. The whole village was bathed in harmony and lived in love. Until the day when, during a horse-back ride, the countess had a fatal fall. Mad with grief, the count disappeared. It was then that misfortunes began raining down on the countryside. Many villagers were forced to go great distances to find work. In dire poverty, families broke up. Claire's stories contrasted the people who stayed on in the area with those who left for town: shepherds and woodcutters versus workers and servants; those who lived on fresh air and freedom versus those who wasted away in a cage, enslaved by work, humiliated, forced to knuckle under "like kids." And then nature, too, got into the act, with violence, even with cruelty. The cruelty of life. The cruelty of animals and humans. In Claire's stories, men fought bears. Babies were abandoned in the mountains. Wolves killed shepherds and ewes. Woodcutters hacked off children's limbs with the blow of an ax. Dogs were found hanged in barns. And, above all, her favorite scene: women thrown from mountaintops to the foot of a cliff.

All these stories made me gasp for breath and shiver with fear and pleasure. Soon I had heard the complete repertory and knew it by heart, but I never wearied of asking for more, waiting for the slight change, the little variation that revived both elation and fright. And then Claire's stories also touched on the world of cats. A secret world different from the one we humans know. Claire had several cats, whose looks, strengths, and weaknesses she would enumerate in detail. By constantly living with them, observing them, making an effort to understand them, she had come to ascertain that cats have, at the tips of their ears and whiskers and in the pads of their paws, sense organs unknown to us. All this gave them access to things we have no inkling of: to a universe on the fringe of our own, but just as real; a whole clandestine world which Claire resolutely believed in.

I have often wondered whether I made up these stories later, but I don't think so. Even today, they baffle me. Summoning up those evenings of madness, picturing that face ablaze with passion, once again my temperature begins to rise, my pulse to quicken. It was then that I had my first feelings of violence and injustice. It was then also that I saw come to a close a stage of my life as a child, like the end of a primitive serenity. After Claire, things were not quite the same, as though with her ceased the perfectly transparent happiness of childhood. For I dimly sensed that all these stories must cover up some deep wound in her; that she found in her imagination a weapon against the disappointments, failures, humiliations of her life. Later, I learned that Claire had had a child by a man who abandoned her. Her father had thrown her out. Her child was dead. In short, nothing but the ordinary.

I learned to read early on. It was mainly my mother who taught me. First with large letters, each drawn on a square that showed objects with names beginning with a particular letter. Later, she got into the habit of answering some of my questions in writing, word by word, in capital letters. Then, she had me read in the street, syllable by syllable: posters, signs, names above shop fronts: ME-TRO-PO-LI-TAIN; E-PI-QUE-RI. Then came books. Soon I could read. I read everything I saw, everything written that caught my eye. And now I am unable to avoid reading. I have become obsessed by reading. I am a maniac for words. A man sick with the written word. I read everything I see: on walls, posters, shop fronts, newspapers over people's shoulders, parcels, labels, packaging, prospectuses, tickets, bills, road signs, want ads, billboards, stamps; in short, everything made up of letters. The written word fascinates me. I am attracted to it as a moth to the light. A billboard on the landscape makes the landscape vanish: I read the billboard. When someone is talking to me, I see the written words file by. For me, a thing is first of all the word with its sequence of letters, its spelling. Often, the very manner of writing, black on white, gives the word, and, thus, the thing, a particular quality, a quickness or a stillness, a lightness or a heaviness, agility or awkwardness. There is an abyss between a *porc* (pig) and a *port* (port): *porc* is extended by the *c* which bends it into round shapes; *port* is tightened by the *t* which stands erect like a crane on a dock. As for *pore*, its *e* makes it longer, perforates it, lets it exude. If someone says *perroquet*, if I think *perroquet*, I immediately see the letters parade by: the *p* pops, the *r*'s roll, the *q* clicks. Only later does the parrot itself come to mind. Nothing sets me dreaming like the sentences in a book, the words written in it, what I see behind the words.

I began learning to count and later to do arithmetic with my grandfather, Albert Franck. I revered him. He was my ideal, the model whom I tried to imitate in all things. From a rather poor family in Lorraine, he went to the Ecole Polytechnique. Becoming an artillery officer, he served in the First World War in the Orient with General Maurice Sarailh. After the war, he was the first Jew to attain the rank of four-star general. Of medium height, with broad athletic shoulders, he had an upturned mustache and a tuft of hair beneath his lower lip. He had both strength and gentleness, a rare mixture of benevolence and authority, of tenderness and energy. He was my rock. It was largely on him that I relied as a child to build my representation of a coherent world. I could ask him about anything, no matter what. He always answered with what I knew was rigor and honesty from the day when, to one of my questions, he replied, "I don't know, but I'll find out. I'll have the answer for you tomorrow." That an adult, a grown-up, could not know, could hesitate—better, that he could recognize that he didn't know, but would find out before answering—this jarred many of my beliefs. It was a revelation, a door opening onto a new world. From that time I had absolute trust and confidence in my grandfather.

For his part, he showed infinite patience with me. When I was seven or eight, my handwriting was illegible. Then he decided to remedy it by taking advantage of a summer visit to Etretat. I had a horror of schoolwork during vacation. I worked hard in school. So I saw no point in having more of it. Everyone agreed with me about this, especially my grandfather. He arrived in Etretat with a complete set of writing materials: quill pens, rulers, colored inks, steel-nibbed pens. And also with a beautiful book of reproductions of sketches by the old masters: Rembrandt, Dürer, Goya. He first had me note what could be drawn with the stroke of a pen: how the stroke's curve could be changed, the fineness or the grain modified. Then both of us tried drawing with a pen some simple object, an apple, a tree, and playing with the stroke, scratching the paper or skimming it, clawing it or caressing it. My grandfather then showed me that writing could be just as much fun. He used some violet ink that, when dried, had metallic green glints. I admired the lines he traced: his handwriting was calm, solid, modest, warm, generous, confidence inspiring. He showed me how to change my handwriting style, make it simple or complex, sad or happy, friendly or aggressive. In short, he showed me the rules of a new game. When he left at the end of two weeks, he had succeeded in transforming my handwriting: molding my letters, sculpting downstrokes and upstrokes, had become an almost sensual pleasure.

It was in Le Mans, where he commanded the military area, that I saw my grandfather at the pinnacle of his glory. Several times my grandmother took me to see a parade on the town square. And when, the troops presenting arms, the band playing "That is the general passing by . . . ," I saw my grandfather, in parade dress and oak-leaved képi, arrive on horseback followed by his ordnance officer, proceed to salute the flags, and finally come to attention during *La Marseillaise*, I was proud as Punch. That settled it. I would be a student at the Ecole Polytechnique, as my grandfather wanted. I would become a military man. My grandmother had knitted me a sky-blue sweater on which she had sewn all the accessories of the general's uniform: buttons, stars, epaulets, collar patches, decorations. In this outfit I oversaw many inspections, led many a parade, joined many battles in my room.

I saw some of my first theatrical presentations with my grandparents. My first film was *Ben Hur*, and afterward I spent many afternoons driving a Roman chariot or freeing galley slaves who, chained to their benches, were rowing under the slave driver's whip. My first musical comedy also: a story that took place in Corsica and concerned Napoleon.

> Corsica's a charming country,
> It's a true paradise . . .

In this, I discovered Napoleon and Bonaparte to be one and the same person, a character I often impersonated in my games. Though my grandfather was amused by this childhood passion for Napoleon, he always attempted to moderate it. He would take me for a walk, my hand in his, to explain *la patrie*. He described what the army must be: its role, its duties. He said this with neither passion nor great phrases. He spoke on these questions simply and precisely, as though it were a problem in mathematics. For him, force used for oppression was detestable, but force used for defense was necessary. He told about the German invasions and his memories of the war. In his eyes, a soldier's greatness was to fight for others, to protect the weak, to prevent violence and extortionate demands. Though Napoleon had been a great war leader, one of the greatest in all history, he had not known when to stop. But, added my grandfather, squeezing my hand, rare are the conquerors who stop in time, who do not always want more: more power, more honors, more money.

With him, army, state, institutions, all could be fitted into place as in a jigsaw puzzle. Each person had his role. The whole transcended the individual. Under his influence, the image I formed of a nation was of a harmony grounded not in divine will, but in the will of human beings. In this whole, each element had its function, its part to play, as in an orchestra; everything was important. And, in fact, my grandfather was interested in everything. An excellent mathematician, he was excited about the revolution in physics and in astronomy. He had played the flute in the Polytechnique's chamber orchestra and was a fencing champion. A great connoisseur of painting, he was interested in history and literature. He selected the books he gave me with as much care as he put into the choice of cheeses and wines for the family's vacation meals: first the classics; then Victor Hugo, Alexandre Dumas, George Sand, Jules Verne, adventure stories from my favorite collection; and then mythology, the Middle Ages, Greek or Roman tales and legends. All this folklore from antiquity that impressed me so strongly. I suspect now that my grandfather had a secret taste for Greek goddesses and patrician Roman women, a taste he transmitted to me.

As I have mentioned, my grandmother had Juno's haughty bearing. She and my grandfather adored each other, but she was not easy to live with. I remember one vacation morning in Dijon when, because the house was filled with visitors, a folding cot had been set up for me in my grandparents' room, at the foot of their bed. I still smell the tea and toast my grandfather brought in for breakfast for the three of us. And that was the moment for long chats about the events of the previous day and about plans for the new one. One morning I was awakened earlier than usual by a peculiar sound. It was my grandmother speaking softly, but apparently in anger. I didn't stir. Heart racing, I kept my eyes closed and my breathing regular, pretending to be asleep. Furious, my grandmother went on in a hissing tone, in fits and starts of curt phrases. She was sharply attacking my grandfather about some business I didn't grasp: a dispute with cousins in which she was criticizing my grandfather for his passivity, she even said cowardice. My grandfather didn't respond, but I heard his breathing quicken. I would very much have liked to see their expressions. But I couldn't risk it, as I would have had to turn around and sit up in bed. My grandmother continued to hurl abuse on my grandfather, until he, worn out, got out of bed and left the room. I heard him shut the door and go out in the street. I resented my grandmother for her hardness, her shrewishness. That one could love someone so much and insult him so much: that confused me.

At seven I started school, in the primary school of the Lycée Carnot. My first impression was definitive and lasted for the whole ten years I spent there, right up to graduation: it was a cage. A cage whose occupants' fortunes depended on the good will of the teachers or proctors who trained them, and were regulated by the sounding of a bell. My mother took me every morning. Although my heart was heavy after she left, I managed not to cry. She came to fetch me back home. Although in the beginning I would hurry to the exit to find her, soon I asked her not to come to the door of the lycée, but to wait for me at the corner of the rue Cardinet. During my years in the lower school, I had two teachers in succession: Monsieur B. had pink and white skin, and a grizzled beard so short that he looked as though he'd made a poor job of shaving; Monsieur S. had not only no beard, but no other hair either. Of these two, I have only rather a neutral memory. Their job was to drill. They drilled. They drilled from competition to competition: tests, honor rolls, congratulations, good grades, prizes. Almost every day one had to leave the starting line to arrive, before the others, at the finish. Or say, before the others did, which city was designated on the unmarked map. Or recite, better than the others, *The Wolf and the Lamb*, a La Fontaine fable. Whether I was more sensitive than my classmates to the carrot and the stick, or showed greater pride or conformity, the class competitions became my obsession, my *idée fixe*. I worked only for them. I lived only for them, encouraged by parents who had an immoderate taste for their offspring's success, one more ransom paid by only children on whom parents focus their dreams and their failed hopes. Competition took such an important place in my life, it was such a preoccupation, that it became the very theme of my games. I was excited only by toys for races or contests. For hours, on little mechanical race courses, I ran horses and metal cars. I named the entrants for my classmates, the favorite being, of course, the one named François Jacob. My father, who took a great interest in the auctions at the Salle Drouot, received catalogues announcing the sales of pictures. Afterward, he would give me these catalogues, which often contained reproductions of the canvases for sale. In my hands, the pages of the catalogues became exercises turned in, at the end of their painting class, by my pupils: Renoir, Pissarro, and Fragonard. I corrected these exercises with red ink, extremely carefully. I graded on a scale of ten both the subject treated and the manner of treating it. And then, with a serious air, the papers under my arm, I entered my bedroom to announce the results to my pupils.

In the elementary grades, I had comrades but no friends. There were a few cliques; for playing war, Indians, cops and robbers. Although I played with the others, I refused to belong to a clique, feeling like neither a leader nor a follower. Until seventh grade, the classes were coeducational. I can still picture a few of the girls: Jeanine, a little sharp blonde with a small thin face, very bright, often head of the class; Françoise, the prettiest, a sturdy brunette who would have attracted me if she had not been too reserved for my shyness to overcome; Tamara, who was Russian and a bit older than the others, with a turned-up nose in a little monkey face, dark skin, straight black hair gathered up on either side of her head in two ponytails. Tamara took an interest in me. She was the only girl who didn't stick to hopscotch, puss in the corner, and tag, but joined with the boys in rowdier games. Several times, on Thursday or Sunday, Tamara came to our house or I went to hers. And there was one afternoon at her place I am unlikely to forget. We were playing in her room, whose door she used to lock: "so Mother won't bother us," she explained. That day we were continuing a game we had begun in the lycée, a sort of war between savage people. Tamara proposed new rules that would deliver the vanquished to the mercy of the conqueror, submitting the former to every punishment and torture invented by the latter. Having downed her on the first try, I had tied her to the radiator with her jump rope and condemned her to look on helplessly while I slowly consumed a chocolate bar. In the second round, we soon found ourselves on the floor. Suddenly Tamara clenched her jaws and, lashing out, pinned me down. It was my turn to be tied with my back to the radiator. She then began a sort of dance around the room, going over to fix her hair before the mirror, coming back to examine me and promise me a particularly severe punishment, railing at me, making fun of me, threatening to call her mother to show off her prisoner. Powerless in my bonds, I submitted in silence to all these outrages. Tamara then approached in little steps. She looked me in the eyes, a slight smile on her lips, and began slowly to unbutton my shirt. Too upset to speak, I didn't know what attitude to take. I shuddered each time she touched me. When she had gotten my shirt off, she drew back to judge the effect: "The punishment has only begun." She went and sat on a chair for a few moments, looking at me with her eyes half closed. I felt both terrified and overwhelmed to find myself half-naked, given over, bound hand and foot, to the will of another. She then came slowly toward me, tickled me for some moments with an umbrella, and unbuckled my belt. Then, with grimaces, smiles, insults uttered in a low voice, and always looking at me straight in the eye, Tamara began to unbutton my pants and to lower them slowly, threatening me with worse treatment. Silent, panic-stricken, unable to move, I felt shame and fear well up: shame of being naked against the radiator; fear of what she could still dream up; fear also of the dark eyes that would not look away from me. Invaded, besieged, I was a prisoner more of that gaze than of the jump rope. And when she reached down to unbutton my underpants, I could not stop myself from crying out in distress, from asking for mercy: "Stop or I'll scream!" She hesitated a second. Then, with a scornful smile, she sneered, "You can be so dumb!" before undoing the cord.

That evening on my way home, I dimly felt that I had put my foot into an unknown world, a strange world that was awakening around me. The memory of Tamara's expression made me suspect secrets that I could not yet comprehend: a different existence from mine, reserved for people more grown-up than I. I guessed at a new use for my body: pleasures that went well beyond the games we had just indulged in; unknown allure of women who might behave quite differently from a mother, an aunt, or a cousin, Tamara's dark eyes would, I knew, go on haunting my dreams. I knew also that a whole other life was waiting to be discovered. A life full of mysteries, pain, unexpected joy. A life whose prospect both frightened and enticed me.

In fourth grade, there was a pupil who elicited everyone's sympathy. Short, skinny, his head too big, his eyes too wide, his clothes too small, Antoine looked like a forlorn puppy. He had lost both mother and father in an accident when he was very young. Since then he lived with an aunt whom he described as the epitome of strictness, severity, and bad temper. What made Antoine's fate so distressing in my eyes was more his life with the aunt than his parents' death. For death did not yet mean a great deal to me. Death was little more than a word. A hoarse and harsh word. A word to avoid. A word representing Evil and Misfortune. A black word, like the black of the armband and ties of mourning, of great hearses pulled by plumed and absurdly decorated horses; the black of funeral hangings adorned with a white letter that appeared overnight over a gateway.

My little pal Antoine had a passion: he loved flies. During recess, he often took me to a corner under the stairs "to show me." He then took out of his pocket a fly cage he had contrived with two round slices of cork fastened together with pins for bars. In the cage, several flies fluttered from bar to bar. With his thin, dirty fingers, Antoine raised a pin to get a fly. He was trying to "figure out how a fly works." He wanted to convince me that the legs were inserted into the body and the wings simply pasted on.

To persuade me of this truth, he began, as he put it, to "disassemble" a fly. He plucked out each leg like a hair. The wings, however, he removed with great care in order to unglue them without tearing them. Who wouldn't get excited about such a mechanism? Since we were unable to put the pieces back together and "reassemble the system," we confined ourselves to observing, in what remained of the fly, the quivering that gradually decreased, like the swinging of a rope that has been shaken. I waited for the gradual onset of immobility.

In the same way, I watched the gradual onset of immobility in chickens that were slaughtered behind the house in Etretat. The operation was usually performed by Jeanne, a neighbor's daughter who came to do the cooking. She sat in a chair, the bird between her thighs. In struggling, the chicken made her skirt come up, and I was fascinated by the sight of the chicken held in those muscular legs. Each time I felt an odd sensation upon seeing that smooth white skin in contact with the silken ocherous feathers, upon imagining the mixture of sweetness and repugnance the young woman must feel when the frantic chicken's fluttering wings touched her thigh. Perhaps the power that emanated from Jeanne had a direct role in my confusion, initiated the beginnings of vertigo brought on by my uneasiness about the scene to follow. Once the chicken was held fast, Jeanne took the neck in one hand and squeezed the head between two fingers to force open its beak. With the other hand, she jammed a pair of scissors into its beak to cut its throat. To keep the blood from spurting forth, the chicken had to be swung around in the air by its claws. For a while the little body flapped until its eyes slowly clouded over. One summer, a distant cousin from the Lorraine came to spend a few days. Anxious to make himself useful, he decided to kill a chicken, but in a very different way. With a blow of his hatchet, he chopped off the head of the chicken, which struggled so violently that it slipped out of his hands. The poor cousin then chased after the headless body, which ran about furiously beating its wings and wound up in a bush. To retrieve it, my cousin had to follow the spots of blood. The chicken was no longer moving.

The death of a fly killed for fun, or of a chicken for dinner, was of small concern to me. Hardly more than the death of animals sacrificed during the Trojan War so that the priest Calchas could examine their entrails and forecast the future. Or the death of a warrior slain in the battle of Thermopylae. I knew that human beings, unlike the gods of Olympus, were mortal. I knew that I was human. But that was only abstract knowledge, with no particular bearing on my own future. My first questioning of death came to me on a warm summer day in that same garden in Etretat. Leading my cavalry regiment, I had decided to attack the enemy with a large flanking movement that would enable me to surround the battalions of red and black currant bushes at the far end of the lawn. Silently, my hand gripping the air rifle I had received on my birthday, I crawled on the grass toward the chestnut tree that sheltered the enemy artillery. Mixed with the whistling bullets and exploding shells was the buzzing of insects, which were numerous at that time of day. Nothing had slowed our advance until from the chestnut tree came the song of a thrush. A thrush that obviously was an enemy observer. My men did not let themselves be taken in by this maneuver. Since my arrival, I had already used my rifle to flush more than twenty thrushes and sparrows waiting in ambush in the trees. As soon as I located the thrush, I put a lead shot in the rifle. I cocked and shouldered it. I slowly took aim. I squeezed the trigger. And the unimaginable happened. The birdsong stopped. Something fell from the branches of the chestnut tree. Going over to it, I found a little heap of feathers, already inert but still warm, the beak contorted, the eyes open, a drop of blood in its anus. The first astonishment over, I felt tremendously proud of my marksmanship. But very quickly pride gave way to shame. Shame at having destroyed something I loved, which played, and sang, and flew, enlivening the garden and the house. And at having done it for nothing. For fun, for the hell of it. I experienced guilt of a new

47

kind, a guilt that had nothing to do with my father or mother or any other human being: the feeling of having interfered with the general order of the world; of having defied an unknown force encompassing heaven and earth, creatures and things; of having perhaps challenged the power I called God. For once, my anguish did not yield to my mother's caresses. That night she helped me bury the thrush at the foot of the chestnut tree. On the grave we set a diamond-shaped stone, and I visited it every year up to the war, whenever I returned to Etretat on vacation. Calling this story to mind today, I still feel a touch of queasiness.

A few days later, at Etretat, a telegram brought word of the death of my great-grandmother. I had seen this great-grandmother from year to year during our holidays in Dijon. I had known her only as a very old woman. So old that there was little hint left in her of what she might once have been. She seemed never to have been young, or even able to have been young. When the telegram arrived, my mother began to cry. Then she pulled herself together to explain that I would never see my great-grandmother again. I was sad to find my mother in tears. But to understand what had happened to my old great-grandmother I had to think of the thrush: that was my model for imagining my great-grandmother's fate.

It was also in Etretat that I saw my first dead person. It was a stormy day, a red flag signaling a ban on swimming. On the beach in the wind, a small crowd, whispering, talking, awaited the return of the lifeboat. An American woman had been swimming despite the ban. The undertow had pulled her out toward the cliff and the striking rock formation known as the Needle. They found her body several hours later. I can still see my mother trying to hold me back, to keep me from going to look; and me breaking away, running over the shingle beach to the boat, coming up just in time to see the removal of a woman's body in a white swimsuit: a stiff, bluish body, bloated from having been so long in the water.

That day, I had begun to see death. For the first time, I had not just heard about death; I had all but seen it at work. But neither my great-grandmother's death nor that of the American woman could touch me directly. My great-grandmother survived herself, so to speak: nothing else could have happened to her than to die. As for the fate of the American, well, it was just an accident. Her end was opposed to true death, the internal, natural death whose progress is undistorted by chance or by men. And to judge by the buzzing of the crowd around the boat, this drowning was the product of more than mere chance. The American woman had been disobedient. She had transgressed, violated the ban on swimming. It was all but said that it served her right, that she got what she deserved.

I was more deeply affected by the death of André, a comrade in the fourth grade. He was a colorless little boy without much personality. There was no special bond between us, just scholastic coexistence. But there was class solidarity, a force linking the pupils. In class, André was seated two rows in front of me. His shock of mussed, washed-out, curly blond hair bobbed up ahead, was a part of the landscape. One morning there was no blond head of hair. Meningitis was mentioned. A few days later, the teacher announced the unexpected death of André; we rose for a minute of silence, as on 11 November, Armistice Day, for the dead of the Great War. Abruptly, death had entered the classroom. It was there, lurking among us. Its presence, its arrogance were confirmed by the empty seat, the absence of the shock of curly hair. It came as a blow to know that death was no longer reserved for the old, that youth, childhood were no guarantee against it: a blow hard to bear. Once again, I had to appeal to the clandestine, to have recourse to the antechamber of life. I had trouble imagining that a boy so young, so normal, so necessary to the group formed by the pupils, could really die; that he could be buried in the ground like the thrush. It seemed to me that André had been withdrawn from the open, official life of the everyday; that he retained a secret life, a life of the night, during which he returned in secret to see his parents. In the morning on arising or during the day, I scarcely credited these beliefs. They crystallized only toward dusk and imposed themselves once night had fallen. It was better not to penetrate this mystery. The attempt might turn about badly, even destroy the very thing I was seeking. To invent death, to construct the idea that sooner or later one dies, was already difficult when it concerned others. That it could happen to me: this did not even occur to me.

The Jewish high holidays of my childhood had a special flavor. In the morning there was some feverishness in the household routine, a bit more bustle, a bit more impatience, a bit louder talk. And when, spiffy in my new suit, my cap on my head, I arrived with my parents at the synagogue, I felt myself the equal of the adults. I felt as though I was coming to a historic gathering, participating in what was of the highest value in the grown-ups' eyes, and able finally to play a role in one of the events that decide the future of the world. To no one would I have ceded the honor of entering that cold and majestic edifice, of passing under those high vaults, yet so gray, so sad, so bereft of artistic elegance or even of religious spirit. Once settled, my legs swinging, in a seat near my father, amidst the hum of the chattering and praying crowd, I swelled with pride and pleasure: pride at being with the men, away from the women whom, by turning around, I could see in the rear, up on the balcony; pleasure in the ceremony and its mystery, in the little flames flickering in the chandeliers, in the old Jewish chants intoned by the cantor's bass backed up by a choir. At the start, everything about the spectacle entranced me: the ushers' outfits, the rabbis' robes, the hats, the beards, the white prayer shawls that swathed the shoulders, the Tablets of the Law unfurled for the faithful. But for me, little by little the spectacle shifted. It came off the stage, with the rabbis and the Tablets of the Law, down into the hall itself, among the seats, around me. Surprised, I scrutinized my neighbors: men with white shawls around their necks and hats on their heads, moving like demons, making sounds, swaying, mumbling, bowing greetings and reverences, turning around to make a sign to the women, interspersing their prayers with interminable nonsense, and, in short, behaving exactly as we were forbidden to do in school. Soon I was overcome with boredom. A dense, deep, irresistible boredom, at having to stay put doing nothing, unable to play, no one to talk to. Then I became insufferable. I didn't stop fidgeting, yawning, kicking the seat, until, exasperated, my father signaled to my mother to take me home.

On the matter of religion, my father's and my mother's families had differing views. On my father's side, they faithfully followed tradition, observed the rituals, fasted. I can still see all the Jacobs at Passover, gathered round the table laid out according to the rite—the seven-candled Menorah, the unleavened bread, the horseradish, the bitter herbs—and listening to my grandfather read in Hebrew the story of the Exodus. My mother, in contrast, came from one of those French families that remain Jewish without bothering about religion, not even giving it a thought, though by no means wishing to deny their origins. For the Francks, one was Jewish the way one was dark-haired or tall or a native of Burgundy. Apart from my mother, who wished to please my father, no one took the trouble to fast or to go to the synagogue. So at the Jacobs', there was regular prayer; at the Francks', no one bothered. And yet, according to all outward signs, the Francks seemed to do just as well, appearing no less prosperous or content than the Jacobs. Hence, my puzzlement regarding the efficacy of prayer, all the more as the best argument I had in favor of prayer came precisely from the side of the unbelievers. It was an adventure that happened to the father of my maternal grandmother, Abraham Bloc, who was known to the family as *le Grand*. During the Franco-Prussian War of 1870, *le Grand* had one day taken his gun and gone, with three or four buddies from Dijon, to fight the Prussians. The escapade had soon come to an end. One fine day, the little band had found itself surrounded by a German company who got ready to shoot them as guerrillas. Facing the firing squad, *le Grand* had retrieved from the back of his mind some scraps of Hebrew to say a little prayer. Startled, the German warrant officer in charge of the firing squad had signaled his men to stop. He had questioned *le Grand*, stared at him for a long time, then suddenly shouted: "Go away! Clear out!" Leaving without more ado, *le Grand* had vanished. Then he had gone home to Dijon. I heard this story many times, but never knew what became of his friends. I did not dare ask whether they were shot.

How, after the story of *le Grand*, could I doubt the efficacy of prayer? How not believe that it gave access to the forces of the invisible that rule the world? How hesitate to pray? Furthermore, I had several times surprised my father at night in bed, arms stretched out, eyes closed, chanting some words of Hebrew. So I made up a game of prayers in which the word *God* constantly recurred. *My God. Good God. My good God. My good Goodgod.* In the latter, *Goodgod*, there was something of gold, of bold, of blood. Mainly, there was Good. Repeating this word that slipped smoothly over the lips, sucking on it, making it melt in the mouth, one could in the end avoid encountering evil; one could appease the forces of the unseen. I made up prayers for every occasion: for holiday projects or for vacation, for birthday gifts, for exams at school. I kept these short, but precisely tailored prayers for the nighttime when I was alone, the lights out. But form soon triumphed over content. The trick was to hit on a concise formula to solicit God's help in some specific situation. To say it. Then repeat it at top speed. Ten times. A hundred times. Two hundred times. Even a thousand times in particularly difficult situations.

One day when I was obsessed by an approaching math exam that, for some obscure reason, I had decided would determine my destiny, I was sitting on a bench in the Parc Monceau, mechanically practicing throwing pebbles into a trash basket. The bench was a good distance from the basket, so my hits were infrequent. In the middle of this exercise, I suddenly thought of making up a sort of test to forecast the results of the exam. If the next stone went into the basket, I would be first. If not, the only thing left was to hide myself forever. With trembling hand and pounding heart, I tossed the stone. It landed in the basket. I came in first in math. After that, prayer gave way to magic and superstition.

Every occasion could be used for this kind of test, even exercises involving neither skill nor strength, but only chance: to close my eyes for five steps along a sidewalk and find that the left foot had reached the middle of a square of pavement without touching a crack; to see whether the license plate number of a seventh car would be divisible by three; to confirm that, of the next six people walking by, at least one would have a beard. Everything was grist for this game's mill. All signs were useful if you knew how to interpret them. No longer was there question of undertaking the least project without first consulting the auspices.

It was from my father that I learned the social game, the role of laws, the rights of others. He was a man of duty, a man filled with respect, respect for institutions, rules, and law. He hated the French mania for "beating the system" and "getting a free ride," the art of fraud, of getting round the law. I cannot remember ever seeing him board a first-class coach with a second-class ticket or drive over the speed limit. He thought a traffic ticket disgraceful. He paid his taxes in advance and for his purchases with cash. He had no tolerance for debts or transactions under the table. But, above all, my father had respect for people, for others. He wanted neither to disturb nor to incommode. He never tried to go ahead of turn; or sent back a bottle of wine presented for tasting; or berated a waiter for bringing lemon soda instead of the beer he had asked for. In society, he avoided talking too loud, imposing his opinion at any price, putting himself too forward. In this, there was no excessive humility or flabby character or timidity, for he never hesitated to place blame firmly, even violently, on people he saw treating an inferior with contempt in a effort to humiliate. He hated arrogance and insolence. He detested people who presumed and domineered. He refused to encroach on his neighbor's territory. In short, he recognized everyone's right to existence.

A conformist in religion, my father was scarcely so in politics. From his youth in Nancy, he had had an animosity toward the country gentleman, the well-born, those he called the "cavalry officers of the noble *de*." He had brought back from the war compassion for the human condition and a taste for justice. His fear was anti-Semitism; his *bête noire*, inequality. He approved of growing rich and powerful through work, tenacity, and daring but not through birth or inheritance. My father's France was that of the Revolution: the France that had recognized the right to live of workers, peasants, and Jews. But he found that France still had some distance to go along this path. For him it was necessary to take more from the rich and strong, and give more to the poor and weak. To this vision of things I strongly subscribed. I was ready to take up my sword in the service of justice and equality. Only sometimes, certain contradictions in my father surprised me: the contrast between obedience to laws and the desire to change them, between the taste for tradition and that for revolution, between respect for divine power and criticism of social power. For a long time I thought that if there were rich and poor, strong and weak, it was because God had willed it so. Gradually, in the picture I was forming of the world, the idea was growing that there exist two kinds of laws: those of God, which must be respected without our being able to change them; and those of men, which must also be respected but can be changed.

For my father, this transformation had to take place in an orderly fashion, through democracy, through the vote. My father loved politics, discussions, elections. He went to the synagogue, and he voted for the Left: a Left that was not theoretical, but highly empirical. He was not a reader. Of Marx, of *Das Kapital*, of Fourier, I never heard until much later, at the lycée. My father's Left was that of the socialist Léon Blum and the radical Edouard Herriot, possibly of the communist Maurice Thorez. From hearing him, I came to see the Left and the Right as Good and Evil, and without hesitation I resolutely took the part of Good, and made myself its champion. The only shadow in the picture was that certain members of the family, notably some among the Francks, were on the side of Satan. At vacation meals in Dijon, my father, who was sure of himself on this subject, would tease my grandfather and particularly my uncle on their political leanings, until my grandmother, with an imperious wave and a sharp

"If you please!" stopped all discussion cold. After the meal my uncle and my father cleared out to proceed with their debate. In my mother's family, however, my father had an ally in my uncle Henri, who was Herriot's doctor. Politically, they were thick as thieves. I recall election night in 1932 when Uncle Henri had come to dine. We had had a radio, then called the TSF, for only a short time. After dinner came the early returns. I can still see my father and Uncle Henri sitting next to the big box of the TSF, with beer and pretzels on a table. As each leftist victory was announced, Uncle Henri, implacable in his gray beard, his eyes crinkling with satisfaction, had a swig of beer and ate a pretzel.

My father sometimes got angry: violent angers, but short-lived; confined to the house; directed against my mother, or against me, or against both of us. For a long time I accommodated myself to them. My father yelled. Sometimes he shut me up for a few minutes in a broom closet. Rarely did he slap me; never anything of any consequence, until the time of my bar mitzvah when I was twelve or thirteen. I went for instruction to a rabbi, a family friend with a very black beard beneath a very soft, very pale face. There was, among other things, a text in Hebrew to learn by heart for chanting on the day of the ceremony. For me, it was a lesson to be learned like any other, like Latin or history. I often went to study in one of my favorite study nooks, the toilet. I would shut myself up in there for hours on end, studying a page from Shakespeare or softly reciting a scene from Molière. One evening, after memorizing the Hebrew text there, I was surprised by my father as I came out of the little room with the Bible in my hand. Suddenly beside himself, he made a scene of unusual violence on the theme, You do not take God to the toilet! The Bible is sacred! You don't stick the Bible in poo-poo.

As I write this, I ask myself whether this scene really took place, whether it was real, whether I actually lived it, like others that come out of the shadows, reappearing after years of oblivion. There again, the blend of shame and rage I feel welling up even today gives this vision the stamp of authenticity. Beside the clearest images, I find blanks, lacunae in the memory, like a fresco from which whole panels have fallen away and disappeared. Only fragments break free. I still see my father pacing, biting his mustache, with my mother unclear which of us to minister to, trying unsuccessfully to calm my father. And I running to shut myself up in my room to gnaw over my resentment, vent my spleen at injustice, incomprehension, loneliness. To vent, too, my anger, which pounded in my breast like waves and made me stumble and bang full-tilt into the furniture. Far from convincing me, my father's "blowup," the very idea of profanation, made me gnash my teeth, made me burst into laughter. Laughter from fury. Laughter from rage. The same laughter that had seized me some months earlier when I had hurled into the toilets the nice little comrade who had called me a dirty Jew. Religion and cesspool: an association that often returned.

What my father may have been denying in this association was the ailment that had recently begun tormenting him; an ailment that proved to be long, persistent, mysterious. A matter of entrails and viscera, of liver and intestines. While this disorder did not stop him from reaching the age of ninety, it weighed on him all his life, and on the lives of those around him. Uncle Henri was not up to this. So he took my father to consult with the leading figures of the moment: consultations that necessitated long preparation, that proceeded in great solemnity, according to a precise ritual, which always translated into new pills, new drops, a new diet, each time more strict, each time contrary to the one before. In fact, no one understood this ailment. And the pain, fatigue, and uncertainty made my father irascible.

The crisis came shortly after my bar mitzvah, which had gone well. I had not, however, experienced the satisfaction I had expected. I had remained indifferent, unmoved by the atmosphere of celebration. Besides, I had seen then more clearly than before the skepticism, almost irony, of two of the family's stronger personalities: my grandfather Franck, the military man; and my uncle Henri, the doctor. From then on, my father more often would ask me on holy days to accompany him to the synagogue. I felt increasingly allergic to those excesses of fervor that I increasingly noticed: pretenses, pretexts, everything that smacked of hypocrisy and exhibitionism. The next Yom Kippur, I began to fast for the first time. Since at the synagogue the seat next to my father was taken, I found myself next to a tall, thin young man with a black beard and bowler, plunged into a maelstrom of devotion. He sat down, got up, broke into bows and scrapes, from time to time chanted a phrase in Hebrew in a high-pitched tone, and, throughout, continually cast furtive glances left and right to make sure the manifestations of his piety did not go unobserved. Fascinated, I watched him for some time, as I would watch an animal in a zoo. Suddenly a question flashed into my mind: "What if God doesn't exist?" To ask the question was to resolve it. The answer seemed to me plain. Everything—synagogue, rabbis, God, prayer—all this was merely a farce that rolled on from age to age, like an ocean created by the credulity and anguish of men. The heavens were empty. Man was alone. Alone, man did what he could. As best he could.

III

IF GOD DID NOT EXIST, it was necessary to do without him. An empty heaven left an earth to fill, and it was up to me to fill it. A world to construct, and it was up to me to construct it. This impression was further strengthened, shortly before his death, by my grandfather Franck. I saw him for the last time in Dijon during the Christmas holidays. Already ill, he was breathing with difficulty and coughing a lot. Once again, he insisted on taking me to his old friend the bookseller to pick up some books selected by him, chosen with deliberation. Going back to the house, in his long gray gabardine coat and fedora, he walked with difficulty. Abruptly, he began to talk about death and the hereafter. "People are children," he growled. "They have to have beliefs, illusions. The illusion of hope. One can't go on without hope. Still, you have to know what to put it in. People go to a lot of trouble to invest their money, they get information, they don't let themselves be duped. But they'll put their hope in any old thing. There's nothing after death, you know. Nothing." My grandfather paused. He took my hand and looked at me. "There's nothing," he repeated. "Nothing. The void. So my only hope is you. You and the children you'll have." He held my hand with such force, his eyes so full of emotion, that I stood there, my throat constricted, without saying a word.

After his death, I needed time to comprehend what I had lost, to realize what my grandfather had given me, to appreciate how much I owed to him my habits, my tastes, my way of looking at the world. One aspect of the teaching in the lycée bothered me: the compartmentalizing of subjects, the isolation of each discipline. Students went from class to class as though they were exploring an archipelago, as if they were visiting a series of countries each governed by a ruler utterly indifferent to what was happening anywhere else. For an hour we translated Seneca. The next hour we heard an account of the marvels of the steam engine. After that, a lecture on the physical structure of the continents of North and South America. No sooner had we finished lunch than we took up the function of chlorophyll in green plants, and concluded the school day in the company of the witches of *Macbeth*. Nothing connected with anything else; there was no question of seeing some possible relation between history and mathematics, or between the natural sciences and geography. No matter how competent the teacher, how concerned to do a good job, how good at teaching, the idea never occurred to one of going beyond his boundaries, of showing us that the world is a whole, that life is a composite of many things. Each subject remained a closed system. Each discipline worked like a closed circuit, ignoring the other ones. It was up to the students to set about constructing their little universe and finding in it some coherence. To each his own synthesis, if he felt one was needed.

My grandfather Franck had explained at great length the ways of coherence and synthesis. To hear him tell it, the Greeks were interested in mathematics, and Shakespeare's heroes in geography. With my grandfather I had learned to place characters in novels in their historical context; and to find similarities and differences between the concerns of physics and natural history. Behind the diversity of subjects and tastes emerged the possibility of some unity, the beginnings of a coherence. With my grandfather gone, it was up to me to produce my own coherence and my own syntheses.

My two grandfathers died within a few months of each other. For a long time, I had felt I was leaning back against the past as if against a solid stone wall. I felt shored up by prior generations, standing on indestructible pillars sunk into the depths of humanity. With the passing of my grandfathers, I suddenly had the impression of a rift behind me, as though stones of the wall had been removed. I had this sensation again at the death of my mother, and of my grandmothers. When the last one, my father, died, it was as if the final stone had been taken away, the last pillar. I was no longer leaning on anything. I no longer had any point of support. I had to stand on my own two feet. It is perhaps this sense of gradual demolition that constitutes the apprenticeship of death.

The school playground. Dust. Shouts. Sliding. Kicking. Punching. Skinned knees. The soccer ball hitting you right in the face. The little secret revealed on the sly. Torn clothes. Lessons furtively reviewed. And then, the gangs with their leaders and laws, their badges and slogans, their hatreds and their scapegoats. Tenth grade was the school year 1934–35, amid 6 February* and the Popular Front; amid the factions of the Action Française, the Croix-de-Feu, the Socialist Youth, the parades with black or red berets and big sticks. Charles Maurras and Léon Blum. Colonel François de La Rocque, and Maurice Thorez. That whole flood of people roaring through the streets came to blows just outside the lycée, even spilling into the playground.

I had learned to distrust violence and racism. At home, I sensed my father was edgy and on his guard during the February riots.

---

* On 6 February 1934, rightist groups marched on the Chambre de Députés. They were stopped by the police on the Pont de la Concorde. Several people were then killed. Charles Maurras and Colonel François de La Rocque were rightist leaders; Léon Blum was a socialist leader; Maurice Thorez was a communist leader.

He feared and hated what he called the brutal rightist blockheads who had the upper hand and were a menace to the republic. He had greeted with relief the mass countermovement born out of 6 February. I resisted labels, parties, orthodoxies. Yet I suspected that, in times of crisis, it was impossible not to take a position. I sided with what was called the Left, but at the Lycée Carnot, the Left was in the minority and lacking in energy. The bulk of the class remained amorphous, without opinion or political leaning. Only the Right was effective, aggressive, organized, under the aegis of Jean de H. who, repeating the tenth grade, was a militant in the Action Française. Golden-haired, eyes of azure blue, a sneer on his lips, H. prided himself on his mustache, which he meant to be dashing but was so pale and sparse that he had to dye it to make it visible. With the support of five or six sidekicks, most of whom were tenth-grade repeaters as well, he strove to rule the class by terror. At the start of the year, I had tried to discuss, to persuade. Very quickly, I realized that I had not the least chance of influencing the behavior of this group. That all my talk was like beating my head against a stone wall. That, for them, the best of arguments had nothing on the least of punches in the nose. I also soon understood that in no case should I yield to threats and blackmail; that any sign of weakness, any flinching, would ultimately cost me dear. Hence, the repeated brawls, the torn jackets, the black eyes, the bloody noses, that my mother greeted sadly and tenderly, but without reproach. It was neither bravado nor a taste for adventure that led me into these recurrent scuffles. Rarely did I have the upper hand. I even ended up being afraid: afraid of that endless brawl; of fighting single-handed against two or three. For several months I felt, each morning on awakening, a kind of nausea at the prospect of this undercover, sly, incessant warfare. At the propect of having once again, that very day, to hit and to be hit.

During one of these tenth-grade recesses, I found myself pinned against a wall by H. and two of his acolytes. They wanted to make me "confess" that the socialist Léon Blum was a piece of garbage, that he ought to be hanged. I suggested putting the reactionary Charles Maurras on the gallows instead. Hence, the first blow. Backed up against the wall, I kicked out in all directions against these jackals. Suddenly, two of my assailants toppled over, their heads knocked together. Half stunned, they slunk off, giving way to the tall form of Bernard L., who was all smiles at having saved the day. "Want more of the same?" he asked Jean de H. H. smoothed his mustache and retreated in good order. "You don't lose anthing by waiting," he said, as the bell signaled the end of recess. Bernard helped me to recover. Tall, slender, with the tapering face of a rabbit, he had taken a liking to me some months before; a friendship that redoubled from the moment he came to my aid. This was my first contact with the Monsieur Perrichon syndrome,* one that I have encountered many times over the years.

Unlike the child, the adolescent has a past. Not just because of the succession of days he has already lived, the thousands of mornings that have dawned, the evenings that have fallen, the moments of laughter and grief. Not just because of time with its accumulation of desires and fears, of the hours of impatience, the itineraries of boredom and joy, the encounter with others: all this tissue of lethargy and fervor; this voyage into passivity and action; everything that is used up, spent, vanished; that has disappeared without warning, without hope of return. But, rather, through constructions already established; through habits already anchored; through what one has already seen of the world; through tastes acquired, codes learned, positions taken; through the unique organization of a little fragment of the universe that has gradually become isolated, crystallized, defined as identity; through the new need to open out toward others, to exchange, to give as much as to receive. By tenth grade I had friends: Roger D., Michel B., Bernard L., this the friend who so opportunely came to my aid. This group gathered at the end of each class, during

* Monsieur Perrichon is the main character in a play by the nineteenth-century playwright Eugène Labiche. He likes the people for whom he has done a favor, while he dislikes those who have done him a favor.

recess, and when school was over. On the way home in the evening, we dawdled, commenting on the events of the day, hardly going three hundred yards in half an hour. Sometimes we set out without a destination and lost ourselves in the city and its crowds. A big city, when one is fifteen, is a continent to explore, a mysterious territory, with countries, inhabitants, even languages that change from one street to the next. There was a secret life in the city. The possibility of vanishing for an instant in the crowd. Of getting lost in the flow of passersby and the jungle of the streets. Of looking for a smile in the faces of the girls one dared not approach.

The spring in Vichy, where my father was taking the waters. The park, the springs, the hotels, the lush green countryside, the Allier River. At the tennis club, just outside town, I am playing with a man I met this morning: about thirty-five years old, short, stocky, with jet-black hair, pale skin, a slight hump in his back jutting out under his jersey. A game of no great interest to me. He plays poorly, for the exercise and to lose weight. I win the match, though without distinction. Tired, sweaty, we head back to town on foot, walking slowly under the hot midday sun. He asks me about my studies, my family, my interests. Arriving at his hotel, he suggests, to save me waiting, that I accompany him to his room while he changes. In the room he takes off his shirt and shorts. Embarrassed by this man in underpants, sticky with sweat, black hair matted on his chest, spinal column jutting out between his shoulder blades, I stare vaguely out the window. He dries his torso. Eyes wide, he comes over to me. Startled, I back away. He caresses my arm. I back off farther, to the wall. He grips me gently by the shoulders. Then I see his head coming close, with its large drops of sweat, his gaze intent, his breathing hoarse, his dreadful grin. He tries to kiss me. Terrified at this nightmare vision, I push him back and flee. I race down the stairs. I run. I run. Fifty years later, I am running still.

At age fifteen or sixteen, my vacations in Etretat were not confined, as at seven or eight, to the garden of the summer house. I felt ardors sprouting, new curiosities, a need to expand my field of action, to seek out others, to bite harder on life. I tried my hand with girls. Notably with Beryl, a little American who had a wave of blonde hair dipping over one eye, a teasing smile, an alluring bosom, and sometimes showed a glimmer of interest in me.

One of the most popular spots in Etretat was the Lecoeur pastry shop. One day when I had taken Beryl there for a creampuff, an unknown girl came in. At her entrance, the color of the daylight, the very quality of the air seemed to be transformed. Tall, slender, with golden highlights in hair and eyes, a blooming complexion, lips glistening like the flesh of fruit, this girl glowed with life, health, joy. Attached to her skirts, she had in tow two little boys who, roaring with laughter at the stories she was telling them in an undertone, begged: "I'm allowed to have cakes today, Odile." "Me too, Odile. I want two, too, Odile. I want to take one back to maman." And Odile smiled, caressed the boys, kissed them, discussed the merits of the various cakes, chose an éclair for the bigger boy, a fruit tart for the little one, hesitated for herself, pointing the tip of an interrogative tongue between her lips, laughed, whirled around, split a *palmier* between the two boys who were laughing and fighting, joked, wiped a speck of whipped cream off her blouse, spun round again, went back, cast a greedy eye on the rum babas, ended by wolfing down a chocolate éclair with ferocious appetite. A whirlwind of smiles and elegance. A blend of exuberance and sweetness, of tenderness and gravity. Never had I seen the like. A sylph, a fairy.

From then on, I had but one idea: to see her again. For several days, I installed myself in the Lecoeur pastry shop. I didn't budge. I took up residence. I spent my days waiting for her. My eyes riveted on the door, I ate cakes, slowly, more slowly yet, in little mouthfuls, crumb by crumb to make them last, each one, as long as possible. A *palmier*. An éclair. An almond cake. Another éclair. A fruit tart. A creampuff. Another *palmier*. I felt sick. I was on the verge of indigestion. But for nothing in the world would I have left the shop. Alas! she didn't return. Once, from a distance, I saw her in the street. Then no more. She was gone. I did not know who she was or where she lived. All I knew of her was her name: Odile.

And all winter long, I thought of Odile. In my room at night, I recalled her smile, the mass of hair that undulated with each step, that swung to her rhythm. I heard her laughing and playing with the little boys. I murmured her name. No other girl I knew had this name. No other name sang with such music, flowed so mellifluously, glided down my throat with such sweetness, like a slightly sugared liqueur, Benedictine or Cointreau. There was, in this name, liquidity and gravity. There was something of the odalisque and of silk, the idyll and the idol.

All winter long, I kept warm and secret inside me this radiant image of young womanhood, so closely attuned to her name which I spelled out, which I said over and over again in a sort of delirium, while painstakingly concealing it from my parents and friends. With the imminent arrival of summer, it was the prospect not just of vacation that beckoned, but of seeing Odile soon again. For the idea that she might not return to Etretat did not even occur to me. Without the least argument, without the least reason, I was certain of seeing her again. And, indeed, I had been on vacation but a few days when I saw the tall slender figure I was hoping for. Even lovelier than I remembered her, more smiling, more luminous, Odile was wearing a light-colored blouse and a full pleated skirt that, with each step, flared out like the petals of a flower. This time, as soon as I could, I managed to learn her name, to find out who her family was, where she was staying. From then on, the village of Etretat was centered on a single person. Everything that touched on Odile, that formed her frame, her setting, took on, in my eyes, a particular quality, a particular relief. Among a summer population consisting mainly of vacationing Parisians, my imagination was focused on just one family, which it had singled out and ennobled, as it had transformed one of the houses and summer villas into a palace for

Sleeping Beauty. I had discovered that my parents had some acquaintance with Odile's aunt and uncle, at whose house she stayed at Etretat. Thus, it would have been easy to approach her. But that was out of the question. I wished neither to divulge my secret not to expose it to smiles and to irony. I preferred to loiter near Sleeping Beauty's palace. I waited for Odile to go by. But when I encountered her, I did not know what to do or say. Once, when she came by loaded down with parcels, I offered her my services. She refused point-blank. To make her acquaintance, to meet her according to the rules, I had to wait until the variety show.

This show was the high point of the summer, the event of the season: a series of sketches written and performed by the vacationers who, with a mixture of gentility and ferocity, poked fun at themselves to tunes of current popular songs or the operettas of Jacques Offenbach, who once lived in Etretat. Two weeks before the show, the rehearsals, the sewing of costumes, the devising of sets made the casino and its little theater as agitated as bacteria in a test tube. I had never performed in the revue. But as soon as I learned that Odile was taking part in it, I signed on. As fellow troupers, it was easy to be formally introduced. It must be confessed that my role was modest. A lowly carbineer in a squad who sang stirringly:

> We are the carbineers,
> The guards of the estate.
> We are coming, we are coming,
> We always come too late.

Odile, however, was one of the stars of the show. In a floor-length gown of red taffeta with a low-cut neckline, she sang and danced a duet with a tall dark-haired fellow, to a melody from *La Vie Parisienne*, a famous operetta by Offenbach. Seeing her glide into the arms of this dandy with slicked-down hair, I felt a wave of savage jealousy sweep over me. Still, I had achieved my aim.

After rehearsal, I was allowed to bring Odile home in the evening. I was allowed to fetch her, to walk and swim with her. I was allowed to help her learn her part, to slip up close to her in the darkened hall during rehearsals.

Every night I went to sleep among the memories of the day. Every morning I awoke filled with joy at the prospect of seeing Odile again. Without fear of comments from my parents, I chanted hymns to the sun, to the sea, to life, so happy was I to rejoin her. I decked myself out for her. I dressed for her. Waiting to see her again; blissfully waiting all day, from morning on. To know that she would be there, would go out with me, would speak to me. I wanted neither to go to her uncle and aunt's where she stayed, nor to have her come to our house. I wanted to keep my secret. My mother was no fool. She said nothing, but had sometimes a different way of looking at me, with a little smile of a new tenderness. I would have loved to see them together, Odile and my mother. I was certain they would get along famously. They went so well together. Without ever mentioning her name, I seized on every chance to maneuver my parents into talking about Odile's family. I repeated this name over and over in a low voice. But it had ceased to seem entirely innocent. The pleasure I took in saying it, in hearing it, had taken on such a flavor of clandestinity, almost of guilt, that my thoughts, my hopes seemed to me discernible by all. I dared not speak of Odile but, while dreading it, wished with all my might that someone would talk to me about her. I ended up using little subterfuges which I found ridiculous but could not do without. At the slightest opportunity, for example, I would propose to my friends the following riddle: my first is a bird; my second is a fish; my third is a piece of land surrounded by water on all sides; and my whole is a reptile with legs: thus, "crow" + "cod" + "isle" = "crocodile." And in public, indulging in the same stupid word game, I would say the first syllable of the riddle's answer, "Croque," to summon up its last: "Odile."

Odile was to leave Etretat three days after the show. For her final day, she agreed to go for a walk alone with me. That morning, barely awake, I ran to open the shutters to see whether the weather was fine. Yes, it was a beautiful day. Yes, it was going to be warm and sunny, with a blue sky and the song of birds. An hour before the rendezvous, I was already hovering around Odile's house. When she appeared, sparkling in her blue gingham dress and espadrilles, not for an instant did I doubt that I was going out with the loveliest girl in the world. We had decided to have a snack at a farm, going by way of the cliff and the seaside to return via an inland route and the Criquetot road. After lunch, the sun at its height, it was a tough climb up the cliff. As we clambered up, the sea receded, and the lapping of the waves against the cliff became fainter. At the top, near the monument to Charles Nungesser and François Coli (two aviators lost while trying to fly over the Atlantic in 1927), there blew an ocean breeze, warm and cool at once. From above, a breathtaking panorama, thousands of times admired, thousands of times renewed by the quality of the air and the vibration of the light. With, that day, the intense blue of the sea from which sunbeams were radiating, the pale blue of a cloudless sky, the chalky gray of the cliff and of the Needle on the other side of the beach, the small village of Etretat spreading out with its black and red roofs, the stippled line of swimmers near the beach, two ships slowly slipping beneath the horizon. And the concert of seagulls: now standing, wings quivering; now veering straight out from the face of the cliff; now diving for some prey. Against this background stood Odile, with her sweet oval face, the full-lipped tenderness of her smile. It was a wondrous promenade to the ends of the earth, in the air of the open sea with the smell of the ocean. We had so much to say to each other: about our studies; our reading; the theater; the people of Etretat; Paris and the friends we had in common. Odile had an astonishing gift for mimicry, a rare talent for spotting oddities and absurdities. I didn't know what I admired more in her: her manner of telling stories, of making fun of people, yet never lacking tenderness; or her grace, that mastery of gesture that comes from complete suppleness and permits one to execute a particular movement without hesitation and with perfect accuracy. Several times during the ascent, I offered Odile a hand to get over a hedge. Each time she refused. Was it reserve? Fear? Disdain? The first refusal surprised me. The third irritated me. I tried to take her arm. She withdrew it. I at once decided that I *had* to kiss her before the walk was over. Immediately the day clouded over. The notion of a duty to perform; the idea of a battle to undertake; the possibility of not achieving my aim; the feeling of incapacity in case of failure: all this snatched away every pleasure from me on the spot. Euphoria was blotted out by anxiety.

From that moment on, I had only one idea: to find the best way, the best moment to kiss Odile. But the kiss had lost its savor. Far from relishing the thought of my lips' delight in touching hers, the idea of the struggle to persuade became a torment. I saw Odile no longer as a haven of sweetness and warmth but as the stake in a conflict, an objective to attain whatever the cost. She was singing as she walked, picking up a daisy here, a buttercup there, doing a little dance step. But this vitality, this joy in living that had captivated me an instant before now seemed merely to increase my nervousness. I thought only of the obligation to kiss her. I'll do it at the next apple tree. No, I'll wait until that hedge over there. And we arrived at the farm without my having tried anything. Under a linden tree, we were served cream cheese, country bread, and cider. Odile ate heartily. I couldn't swallow a thing. Then we had to leave. And with the questions of Odile, who didn't understand why I had lost my gaiety, my misery increased. After a field of alfalfa, the path ran along a hedge of hawthorn before coming to a copse of oaks visible in the distance. I decided to go into action when we reached the woods. Secretly, however, I hoped that some monstrous incident—a fire, an earthquake, the apparition of a raging bull or rabid dog— would intervene to scuttle my plan. As we approached the woods, Odile seemed more and more intimidating, inaccessible. And no fire came to my aid. No wild beast loomed forth. Only a cow in the field gazed at us sympathetically. I had reached such a pitch of anxiety that I no longer dared even look at Odile. Suddenly, I seized her hand and tried to draw it toward me. She pulled it away. Perhaps I should have forced the issue when she shied away like that. But it was in an almost pleading tone that, gently taking her by the shoulders, I said, "Let me kiss you." She struggled as though grabbed by the devil. She pushed me back, then strode off. I followed in silence. We returned to Etretat. She left the next morning.

Why had Odile rejected me this way? Why had she so forcefully repelled my innocent caress? Perhaps I had tried to go too quickly. Perhaps she didn't want, just before her departure, to grant me what, in her mind, I might have taken as a pact, a promise. Perhaps I was simply deluded in thinking I pleased her. For several days, I secluded myself at home. Impossible, there, to elude the vigilance of my mother. With her insight into things concerning me, alarmed by my sudden dejection, but taking care to keep from speaking about it, she tried, in all her tenderness, to restore to me a taste for living. As formerly, when I was ill, she fixed my favorite dishes. As formerly, she brought me little gifts: books, a mechanical pencil with colored leads for which I had long yearned. As formerly, she came at night, for longer than usual, to sit on my bed to chat. At the end of three days, I condescended to leave the garden. A pilgrimage to the Lecoeur shop was called for. The first person I saw there was Beryl. Seeing me, she had a little smile: "Love affair over already?" I disregarded the sarcasm and offered her a cake. We left together. Beryl's fifteen years had neither the imperial stature nor the grace of Odile's sixteen. She was even a bit too short, a bit too plump for my taste. With her mocking smile, her blonde straight locks covering one eye, and the blue gleam of malice flickering in the other, she had, however, a certain piquancy. And then nothing about her intimidated me. I proposed a walk for the next day: a climb up the cliff, a bite at the farm, coming back by an inland route. She accepted.

At the hour agreed on, I found Beryl all smiles, looking smart in an orange blouse and a green skirt. The ascent up the cliff began again, as hard as on the earlier day. But nothing could put a damper on Beryl's prattle: not the view from the monument to Nungesser and Coli, not the sight of the sea, not the radiance of the sky, not the cry of the seagulls. She held forth incessantly in an odd accent and with unexpected turns of phrase, using a word-for-word translation of English expressions into French. Flushed by the wind, hair in disarray, she laughed at her own mistakes. Upon reaching a low stone wall that we had to climb over, she took my hand for support, and did not let go. At the farm, we had cream cheese, country bread, and cider. As we were leaving, she took my arm. I had, of course, resolved to kiss her when we reached the oak woods. But she was way ahead of me. Before I had made a move, Beryl stopped and drew me to her. I would very much have liked to be able to credit myself with the management of the operation leading to that long and very tender kiss. But it was out of the question. All I could do was to put my arm around her shoulder and lead her to the clearing in the woods. There, we sat leaning against each other. And during a second kiss, as ardent as the first, my hand met no resistance in trying to undo Beryl's blouse. Fingers, clumsy with haste, on the buttons. Shining eyes and sultry looks. Damp smell of her hair. Sweet smooth skin. Warmth of the hard nakedness. A lump in my throat at the sight of my first naked breast. Nipple hardening under my palm. When I tried to extend my caress farther down her belly, she stopped me. When I tried to put her hand behind her back, she struggled. When I tried to immobilize her between my legs, she kicked. Breathless from the effort, pleased with each other, we left hand in hand. Arriving back in Etretat, Beryl said to me kindly, *"Je vous autorise à me prendre encore dehors."** Perplexed, I spent the evening going through my English dictionary.

---

* The word-for-word translation is: "I authorize you to take me again outside," but the word *prendre* has quite a different meaning in French.

Nineteen hundred thirty-six, the year of my first *bachot*.* A year charged with history. Blow upon blow: German troops occupy the Rhineland; the Popular Front comes to power in France; civil war breaks out in Spain. The year 1936: I certainly did not understand everything it already contained, all the promise of unhappiness it portended. Even so, my feeling of malaise began then: a dim impression of slippage, of fissures in what had hitherto seemed like a seamless block, an all-powerful France. The vague, confused presentiment of a storm brewing. A threat. The unforeseeable.

For a long time I had seen things as immutable: not only things in nature, rivers, mountains, forests; but also things produced by men, objects, books, houses. Walking in a street, between two rows of apartment buildings, was a little like traversing a valley between two mountain ridges. It was to parade down eternity. One day, in the Ternes district where my paternal grandparents lived, I found myself in front of a work site where a house was being razed: an ancient hovel all twisted and cracked. Inside, masons were busy with pick and ax. Outside, onlookers, noses in the air, were watching them. The workmen enjoyed breaking things; the onlookers, seeing them broken. This spectacle reminded me of my childhood games, when my father built delicate castles of cards or huge stacks of cubes that at first I was not allowed to touch. Until the moment when I was permitted to throw a ball at the construction, which then flew to pieces. That one could treat a house this way, a real house with walls, roof, floors, revolted me. Why pull down the house? Why not simply add on to it, build next to it? Walls were made to last, like rock.

I probably got from my father my taste for the permanent, the indestructible. His feeling for stability was connected with his work. His interest in the land and farms led him straight to the point where the industry of man is most closely tied to the activ-

* Examination given in two parts, at a one-year interval, at the end of high school. It is necessary to take the *bachot* to go on to college.

ity of nature. Farmers could very well change; alfalfa could very well replace wheat in the fields. Yet beyond the changing generations and their vicissitudes endured the walls and the land, immutable. Despite all of transformations in life since my father's childhood, despite the automobile, the airplane, the radio, there were the farm and the field. From this source also may have come his taste for old things, for antique furniture. There is some eternity in a Louis XV commode.

I had even carried over my respect for things, and my belief in their stability, to institutions: the Constitution, the authorities, the civil service, the army, the legislature, the government, the ministry of justice, the Polytechnique were a little like the Pantheon, the Arch of Triumph, Notre Dame. I had become aware of these institutions without grasping their exact function. In the representation of the world I had built, they underlay the landscape, so to speak. They made the scene hang together. They formed the indestructible framework of our country, of our life. Forged by centuries of history, these institutions seemed to me to be the pure product of human wisdom. I scarcely imagined that better ones could be devised. And especially that their solidity was not foolproof, capable of resisting all those who, from within or without, on the far Right or on the far Left, dreamed of overthrowing them. With its laws and its great men, its industries and its museums, its senate and its university, the country seemed to me like a tall white ship, all sails unfurled, cutting into winds and tides. Men, leaders, could well change. Governments could well veer to one side or the other. The superstructure lasted, unbreakable. I could not conceive of a better or more stable system overall. It was the ground on which we trod, the frame in which we lived.

My father had an odd friend, an old buddy from his regiment, Isidore B., known as Isi. Two or three times a year, Isi would come to the house for dinner. He came always alone and was always the only guest. Of these dinners for four, I have a bizarre memory. I scarcely understood who Isi was or his relation to my parents. He was treated in an entirely special way and given particular consideration. But, in speaking of him, my mother and father had little smiles of irony and anxiety. He would arrive with a large bouquet of roses for my mother but, once settled in an armchair, it was he who had the air of receiving. As he was considered a gourmet, my mother insisted on preparing dishes which he lovingly inhaled before tasting them with the expression of a tomcat. Tall, lean, distinguished, always well dressed; with thin lips, high, ruddy cheekbones, and short thick hair, Isi gazed at the world with an expression as despairing as it was diabolical. Despairing because of the war; because a grave injury, involving long months in the hospital, had forced him to give up his philosophical studies to go into the insurance business, which he refused to discuss. Diabolical because of his contempt for, and aggressiveness toward, conformism and institutions; also because of his taste for boys, which, in my naïveté on the subject, took me a while to discern. He had once flirted with that reactionary veterans' organization, the Croix-de-Feu, without ever joining it. Soon weary of speeches and parades, accusing this group of "goofing around" rather than tackling real problems, he took refuge in a sort of anarchism of the Right.

My father and Uncle Henri had celebrated the victory of the Popular Front with swigs of beer and pretzels, as was fitting. Some weeks later, Isi came to dine at the house. That evening he had a more anxious eye, a more sardonic smile, than usual. After dinner, he launched into a protracted lecture on the theme "We are a nation of fools." My father had taught me to distrust those Jews of the Right who loved order, admired Mussolini, and praised the fascist virtues. His childhood memories, his battles with the "cavalry officers of the noble *de*" and their offspring, made him sniff out from afar any suggestion of brutality, hatred, the hierarchy of money, contempt for man. More than others, he denied the Jews the right to support fascism, to associate themselves with anything giving off the faintest whiff of anti-Semitism: in short, the right to provide others with the "sticks for getting themselves beaten." So I distrusted Isi and his speeches. But, although on my guard and little inclined to swallow his words whole, I could not that evening keep from thinking that his arguments carried some weight. For him, a new situation had just been created by two events: the occupation of the Rhineland by the Nazis, and the accession to power of the Popular Front in France. Each of these two facts was in itself fraught with consequences. But it was their conjunction, a few weeks apart, that Isi took as both symbol and warning. It was in it that he saw a prelude to calamities whose magnitude and violence were as yet incalculable. Snug in his armchair, snifter of cognac in his hand, Isi spoke with a passion that I had not yet seen in him. On the one hand, the Germans, an unhappy people bent on revenge; a violent, bellicose, unscrupulous regime; a rearmament that, stupidly tolerated by the French, was now going to accelerate unstopped, fueled by an economy entirely geared to this end. On the other hand, the French, aged, carefree, bled white by war, believing less and less in themselves; a neglected national defense, undermined by pacifism; a nation weakened by financial scandals, press campaigns, partisan passions which split it in two; and, to crown the whole, the coming to power of an erratic and weak-willed government, directed by ideologues preoccupied more with social problems than with external dangers. For the French not to see the threat increasing each day, concluded Isi, they had to be blind, or mad, or stupid. It was the last hypothesis he favored. And he tossed off the last of his cognac.

During this speech, my father had said very little. But his constant shifting around in his armchair testified to his annoyance and disapproval. Though he rarely agreed with Isi's ideas, he appreciated the man who amused him. That evening, I knew he could not accept any of the views expressed. Isi's condemnation of Hitler did not in my father's eyes justify his friend's warmongering. My father had as much faith as ever in the military virtues of the French. And, far from criticizing the new government, he had more confidence than anyone else in Léon Blum, in his ability, in his courage in overcoming the difficulties of those troubled times. As for me, though Isi's criticisms clashed with my views and feelings about France, I was struck by his indictment of Nazi Germany and his rage at the Reichswehr's crossing of the Rhine, which he called an act of war. I was impressed by his argument that the demilitarization of the Rhineland constituted the only possible safeguard against the threat that Hitler was brandishing on all fronts. But mainly Isi's speech had brought home a basic contradiction, an incompatibility between certain of my most cherished values. How could the patriotism of my childhood, my family, my grandfather, all my forebears, be reconciled with the pacifism, the mild and indeterminate socialism, the solidarity with the Left toward which the influence of my father and many of my friendships inclined me? Though the two tendencies had, up to that point, comfortably coexisted in me, I had a hazy intimation that I would be forced to choose between them. That evening I glimpsed the distance separating morality from politics, noble sentiments from plans of action. I considered that in my political education I had reached a new stage.

I began the next stage with Karl-Heinz, a young Jewish-German refugee looking for work in France. He came once a week to give me German lessons. He was a a little, rotund man of twenty-five or twenty-six, pale of complexion, greasy of skin, mild in aspect, with sharp, anxious eyes behind steel-rimmed glasses. He came in wearing an old, threadbare, too-tight overcoat and a broad hat, which he removed with care so as not to muss his well-brushed hair. He looked something like a capon destined for the dinner table. His air of long suffering was accentuated by a hoarse asthmatic wheeze which left him gasping like a badly swept flue. Even so, behind this air of victim, I soon discovered a discreet but ever-present sense of superiority.

I learned little German with Karl-Heinz: a little grammar, a little vocabulary, a few lines of Heine's *Die Lorelei*. What I wanted from him was not access to German literature but his own history, his life, his escape, especially his views on Nazism. To get him to talk, I had to badger him, trick him, argue with him. Once started, however, he could go on for hours. And for those hours I could listen to that voice lacerated by his labored breathing, by efforts to regain his breath. Karl-Heinz's family belonged to the

Jewish *petite bourgeoisie*. He had lived in Berlin with his father, a well-known lawyer, his mother, and his sister. He himself had begun the study of law. One day, his father was arrested. For no reason. As a Jew. By the Gestapo. Immediately, Karl-Heinz's mother packed him off to Kiel, to stay with trusted friends; from there he went to Denmark; then to France. Since his departure, he had had no news of his family.

In relating his own history, Karl-Heinz maintained a cool, neutral tone. One would have said that he was talking about someone else. But, in describing the Nazis and their excesses, in expressing his outrage, he suddenly changed register. His face, normally pale and wan, gradually took on a flush; his eyes danced behind their glasses; his voice rose in pitch to describe what was going on in Germany. Curiously, though his precise, violent images profoundly impressed me at the time, they have today faded from memory. The reason may be that as, in the years to follow, the horror mounted, other equally violent images and atrocities accumulated, obscuring Karl-Heinz's descriptions. Only the main points emerge: the Reichstag in flames, and the street taken over by jackbooted stormtroopers who, truncheons in hand, were cleansing the Reich, hunting down undesirables. Frenzied scenes of carousing and lunacy in the beer halls where the faithful bawled out Nazi songs in a trance. The liquidation, during the Night of the Long Knives,* of old friends who had not been sufficiently zealous in carrying out orders. The columns of regimented, indoctrinated children who paraded at night by torchlight under the guard of swastikaed men who were teaching them to obey force and its ideology. The great crowds of delirious people at the Sportpalast who, with arms upraised, proclaimed in bursts of *"Sieg Heil"* their devotion to barbarity, to power. The endless harangues of Adolf Hitler who, with his unbearably gross and vulgar voice, mesmerized millions of people by screaming anathemas against the communists, against the Western democ-

---

* During the night of 30 June 1934, the heads of one branch of the Nazi troops, the "Assault Sections," were killed on Hitler's order.

racies, against the international Jewish scum. The huge bonfires at the Berlin Opera and at the university, where students threw the condemned books while ceremoniously declaiming Goebbels's execration of the decadent works. Not alone German books, sneered Karl-Heinz, but French ones, too: Zola, Gide, Proust, into the flames! All those vast pageants in red and black, those fires in the night, the torchlight parades, the flaming torches at the Sportpalast, the blazing mass of books: all that symbolized purification, legitimized brutality, acclaimed deliberate cruelty allied to the power of the new masters. Total submission to the Nazi order was thus based on a blend of enthusiasm and brutalization, of intolerance and savagery. It manifested itself in the hunt for Jews, the destruction of Jewish books, the pillage of Jewish stores, the theft of Jewish property, the beating of Jews in the streets, the arrests of Jews, the burning of synagogues. When Karl-Heinz left Germany, people were beginning to talk in hushed tones about camps guarded by Nazi brutes with dogs. Camps from which no one had returned to say what went on in them.

Far from growing inflamed by this recital, Karl-Heinz's face went back to being pale and set as he spoke. He was gritting his teeth in anger, but a cold anger. In the Germany that taught hate, he had learned hate, a hatred pure and absolute. This monstrous world of violence and contempt would soon burst the boundaries of the Reich. The Nazis thought only of rebuilding Germany's military might. The danger posed by this will to rearm had to be denounced all over the world. The whole world must know what was going on, what plans were afoot. Especially the French. The absurd, insensible, gutless French. The French who would soon reckon with the will to power, feel the Herrenvolk's need to dominate. The French who "were soon going to see, for their turn would come quickly." In the eyes of Karl-Heinz, behind those steel-rimmed glasses, I read hatred: of me; of the French; of the French Jews still at peace, still tranquil, who were doing nothing to halt the monster.

Karl-Heinz came regularly for two months. Then, without warning, he vanished. I never heard of him again.

The weeks before the *bachot* exam had a special feeling: a mixture of feverishness and celebration, of work and sunlight. I was a good pupil, but a grind, which was not the thing to be. The fashion was to have it come easy, to succeed without having to work, without sweating it. One was supposed to act casual, to appear to scorn effort, to let on that one was doing nothing. But as far back as I can remember, I was always studious, even avidly so. My life had been governed by an insurmountable sense of duty, which found in the *bachot* exam a perfect opportunity to exercise itself. But, at the same time, this period of my life had a festive aspect. As on vacation, I set my own pace, not the one dictated by the bell of the lycée. On fine days, I even settled in a corner of the Parc Monceau, with an outline of mathematics or a history text. Then study streamed from the sun.

The examination itself was uneventful. An amphitheater at the Sorbonne. For two days, morning and afternoon, my arrival with a Greek or a Latin dictionary under my arm, a bottle of ink at the end of a string. I sit at a table, between a Jacard, Alexandre, pale and debonair under his red hair, and a Jacotte, Marie, serious and dreamy, with a long black braid down her back. Two weeks later, the results. The jostling as one looks for one's name on the list. Then studying for the oral exam. Back to the Sorbonne, to another amphitheater, stifling hot, full of murmurs and whispering. Scattered around the room, the examiners before whom each candidate must present himself in turn. The exam looks at the same time like a confession and an obstacle course. First hurdle: French. A thin, dry examiner, black of eye and hair, hands me a slim volume from the Larousse collection, open to a page for me to comment on. Benjamin Constant's novel *Adolphe*, worse luck! Fifteen minutes to prepare, while another candidate discourses on Victor Hugo's *Les Châtiments*, the lucky devil. *Adolphe*. I have not read it. I know next to nothing about it. Suddenly I recall that these little books always have critical notes. Slowly I lower the book under the table, on my knees, so as not to be seen by the examiner. The notes are there: on Benjamin Constant; on *Adolphe*. So why not pass as an expert on this literature?

Second hurdle: physics and chemistry. A young teacher, chubby, pink, blond. I wait while an attractive girl is being questioned. As she answers, I notice that she is escorted by two students reading for the Ecole Polytechnique. Installed behind the examiner, facing the girl, they each have a foot in one hand and are writing, on the soles of their shoes, the formulas requested by the examiner. While one writes, the other erases. Very probably, the girl knows hardly any physics. But the two boys are so adroit, and their trick so plainly visible to her, that she puts on a virtuoso performance. Second hurdle. Third. . . . Fifth and final one: history and geography. A little old man, white and wrinkled. While he chats with a colleague, I draw a paper from a basket. Two impossible questions! The old man is too absorbed in his talk to notice me. I refold the paper. I put it back in the basket. I take another one: the first Treaty of Paris, the estuary of the Loire. That's more like it. I get started on the Treaty of Paris. My knowledge exhausted, I stop. He stares at me, astonished. "That's all?" I cast about. "Come now, think hard!" I rack my brains. His forehead creases. I can think of nothing more. Then he bursts out, "The Saar, monsieur! You have forgotten the Saar, monsieur. Germany has just retaken it, monsieur, but the Saar is ours! We must recover it!" He is foaming at the mouth. He is drooling. He is strangled with fury. We go on to geography: the mouth of the Loire. "Tell me, monsieur, what is an estuary?" Taken aback, I sputter: "The area . . . river . . . flows . . . sea." "No, monsieur, an estuary is no longer fresh water. It is not yet salt water. It is *brackish* water!" Befuddled, I stare at him. A crackpot! I have run into a maniac! I go through the rest of the interrogation in a daze. Begun in glory, the exam ends in madness. The candidates leave during the deliberation. It is hot. My mouth is dry, my head aching. The president of the jury announces the results: *Passed. With distinction.* Whew!

Etretat. The sea wall, where I have just rejoined my parents. My father in a temper; he found himself with a group of men and women who wanted to enter the casino. "As usual," he says, "paying their way in, like everyone else. They were refused admittance on the pretext that there were too many of them. That they refused to disperse. That they were not properly dressed for the place! *Properly!*" My father speaks in a low, hard voice, his voice of suppressed anger. "Actually, they are not wanted here because they are on a paid vacation. People who have never been on a vacation in their lives, who are seeing the ocean for the first time. Those rich sons of bitches! The most obstinate is old S." With his head he indicates, among a group at the bar, a fat man in plus-fours, his calves showing to advantage, holding a bulldog on a leash, gesticulating, declaiming volubly to an audience who approves of its leader. "He's looking more and more like his dog," adds my father. "The other day he tried to box the waiter's ears for bringing him an anisette instead of a glass of port and for having the cheek to claim that old S. had indeed ordered an anisette. Those people are forever railing against the communists, but the communists couldn't get themselves better recruiting agents." My father made little distinction between politics and morality. He did not rationalize it. He did not theorize it. Reality appeared to him harsh and simple. Thus, he talked about it with simple words, without guile or self-consciousness. Through my father I began to see the weaknesses of this nation, which I wanted to be strong, united. Here and there, one heard cracking sounds, like the start of a great fissure.

Back in Paris, I had, after much hesitation, finally telephoned Odile. Her friendly greeting had given me hope. I had seen her again. Several times in the course of the winter, we went to-gether to dances. Getting ready for these dates unloosed in me a mixture of anxiety and elation. It was rare indeed that my face was not lit up with some bad acne or that my white shirt was ironed to my liking. But these little nuisances vanished in the pride of taking out Odile, the pleasure of holding her in my arms while we danced. The party over, I accompanied her home. One evening, before separating, I won a brief kiss. It was repeated on every date, as an accepted fact. Later, I pressed her a little, mak-ing the kiss more ardent, more tender. But without going further. Tacitly, the kiss became a sort of ritual that neither of us thought of changing, either to deepen it or omit it. We respected a secret contract. Like two armies encamped in their respective positions, observing each other but avoiding any abrupt movement, any fireworks, that might disturb the hard-won equilibrium.

Summer. Odile returned to spend a few days in Etretat. In the excitement of vacation, a still greater excitement occupied me: to get Odile to love me. Her profile filled my entire horizon. In her smile, I saw unclose the portals of the world. Her voice, her gestures made up the mornings of Etretat, with a pale sky and the raucous cries of the seagulls. And also with a new anxiety: not to leave her, not let other fellows slip between us, to get myself into her good graces, to form a part of her landscape so that, later, for her to conjure up the cliffs, to remember the sun going down on the sea would be also to think of me. Even today I recall the fever these schemes produced in me, the importance I attached to each look, to each smile. It was no longer a matter of vanity, or even of pleasure. It had become a question of hope, of the Promised Land.

The vanity and pleasure returned with an unknown girl. She was tall. She was blonde. She had green eyes. Her name was Paola. I had met her in the Italian Alps where three friends and I were spending Christmas vacation. The only beginning skier, I conscientiously took lessons with an instructor. At the end of three days, my friends judged that I had made enough progress to join them on a downhill run. It was only after we reached the top, before the first slope, that I realized what lay in store for me. It took me close to four hours to cover what took my friends about ten minutes, and expert skiers scarcely more than two. In the face of my ineptitude, my friends showed no qualms about leaving me to my fate. A nightmare! A rough descent, now over bumpy terrain, now on an icy footpath in the clearing of a forest. Spills in hollows, over whose tops I saw the good skiers pass in flashes, not hesitating to jump. A confusion of skis and poles from which I sometimes spent many minutes emerging. In short, an utter fiasco. The one bright spot in my misery: during a fall, when I had once again tumbled into the snow, entangled in my gear, there came rolling upon me and shouting a pink and white body, dressed all in white. That was how I met Paola. We helped each other up as best we could and, laughing, brushed ourselves off. Then each of us went off at his own pace.

I saw Paola only once more, on New Year's Eve. My three friends and I had decided to spend the evening in one of the large local hotels. Seats at table that evening were randomly assigned. And I chanced to find myself seated next to Paola. She wore a white, gold-embroidered blouse and a green Tyrolian skirt with suspenders. She lived in Florence and spoke French with a sing-song accent that made me giddy. This cooing of the *r*, sometimes sweet and muffled, sometimes laughing and triumphant, was exactly the accent that my imagination attributed to Sanseverina, the main feminine character in Stendhal's *La Chartreuse de Parme*. Following an argument, the man who had come to the resort with Paola had abandoned her. She was looking for a new boyfriend, preferably rich and open-handed. But being between what she called "meal-ticket lovers," she had decided to treat herself to a few days' vacation. When she laughed, she stretched back her neck, a neck that was very long and very white. The mass of her hair then streamed along her cheek, gilded with fugitive glints of Titian red. In the distance, my friends were eagerly signaling to me, but I ignored them. I asked Paola to dance. She had a full, firm body, which knew how to be pliant and to adapt easily to its partner, but which could also become imperious and seemed to be surrounding mine. At midnight, the lights went out. With no difficulty I found two warm and consenting lips. A short while later, I followed Paola to her room. Slowly, she slipped her blouse over her head. And when I saw her bare breasts spring up, it was as if a long promise, issuing from the depths of time, had finally, there, come to life.

Back at school for the one-year preparations for the second *bachot* exam. In math class, the start of school took an unusual turn right from the first day, from the first morning, with the greeting of Monsieur S., the mathematics teacher. Hardly were we seated at our desks than he sprang a quiz on us, a quiz my classmates agreed was "a bitch," for it was based on notions that went beyond what we had learned the year before. From nine until noon, our heads still full of vacation, out of the habit of reasoning and calculating, we had to struggle with equations that we had not mastered. All the while, his gray goatee held high, his eyes alert behind steel-rimmed glasses, his attitude complacent, Monsieur S. paced back and forth like a caged bear, stopping from time to time to clear his throat and spit into a handkerchief which he then inspected with interest, resuming his pacing, looking over our shoulders, occasionally venturing a faint smile that spoke volumes about the contempt he already had, by the end of the hour, for some of his new pupils who were surprised by the brutality and the difficulty of the test. At its end, the dejection was universal.

Fortunately, that afternoon was the first class in philosophy, a new subject and the only one anticipated with some impatience. All the more so as the teacher, Monsieur L., was reputed to be eccentric. It was said that at night when his children, aged eight and ten, were asleep, he hung them by their feet or put them in some other strange position, before waking them to get them to describe their dreams. An experimenter, in short. Corpulent, with thinning hair, very blue eyes behind large tortoiseshell glasses, Monsieur L. greeted us with a chilly affability. As we took our places, he tapped with his hand on his desk. Silent, he awaited our silence. Suddenly he pointed to one student at random. "You! What is your name?" "Dupont." "And you?" "Durand." "And you?" "Duval." "Well, Dupont, Durand, and Duval, you will each have two hours of detention next Thursday. It will teach you that I'd rather be thought a swine than a dimwit! Philosophy, messieurs, is first of all a way of reasoning . . ." So intense was the silence that we could hear flies flying. And we heard them flying for at least a week in the philosophy class. But these two classes, one after the other on the first day, left their mark. All the more as, in the way of extravagance and despotism, the other teachers yielded nothing to the first two. All those quirks and manias that accumulated to make life at the lycée unbearable! A world of terror and brutalization: terror of being late, of forgetting some homework, of not knowing a lesson, of being sent to the blackboard, of suddenly facing a written test. As if the function of the lycée were not so much to instruct as to bring the young to heel, to standardize them, to pour them all through the same mold. As if the Republic wanted to reshape its children on a Procrustean bed, to transform them into future soldiers, drilled to obey, to march in step without grousing. I had always in my mind the preparations for the Polytechnique. I liked mathematics and physics and won the prizes in each. I was even enrolled for the following year in a preparatory class in a specialized lycée on the Left Bank. But to prepare for the Polytechnique meant at least two years of an even more draconian regime. This idea began to haunt me. Then it obsessed me. One night I had a bad dream. I found myself in a deep hole that I tried in vain to get out of. Above, bending over me, his hat on his head, his goatee cocked, Monsieur S., the mathematics teacher, examined me with the eyes of a horse dealer sizing up a horse. I struggled. I tried to jump, to grab hold of the sides of the hole. But the earth crumbled, slipped away. I kept falling back. Monsieur S. spit into his handkerchief, examined it at length, then, with a little smile, declared softly, "You can never get out." On awakening, I decided not to prepare for the Polytechnique.

This plan had been inspired by my grandfather Franck; and while he was still alive, I would probably neither have wished nor dared to change it. My parents had little reason to object to my decision. The main problem was to determine what to do instead of the Polytechnique. I had no sense of vocation. Just a slight interest in medicine: perhaps owing to my admiration for the personality of my uncle Henri; perhaps also because one of my friends constantly boasted about the surgical prowess of his father. Surgery must give you a sense of being useful, of serving your neighbor. There was an 'teresting side to the profession of surgeon, an aspect of sorcerer, even of demiurge, that tempted me. I decided to consult Uncle Henri. For the occasion he invited me to lunch with him at one of his favorite restaurants, near the Opéra. As we entered, the waiters hurried over to greet the "doctor" and escort him to his table. While smoothing his beard, his glasses pushed up on his head, Uncle Henri seemed to be caressing the menu with his eyes. After much deliberation, he selected three dozen oysters while forbidding me to have any at all: "You haven't been vaccinated against typhoid." After which he wanted me to have, like him, a stuffed cabbage. I dared to say that I didn't care for cabbage. "You're wrong. You don't know what you're missing." After which, snug in his chair, the napkin tied around his neck, he attacked his thirty-six oysters, washed down with a bottle of Sancerre. Then, and only then, was it possible to talk business with him. And, to begin with, Uncle Henri strongly advised against going into medicine. It wasn't what it had been in his day. Patients had become demanding. They pestered you for a yes or a no. You had to kill yourself just to make a living. In short, it had become a dog's trade. For it to be monetarily interesting, one now had to be a hospital physician and professor. That meant taking the competitive exams up to the age of fifty. And then there was no time left for living. So one had to be either mad or stupid to choose such a life. As for surgery, it was even worse. Medicine required a certain flexibility of mind, a certain finesse. In surgery, not even that. You cut. You take out. You sew back up. On that point, having finished his stuffed cabbage and the last of his pear Melba, Uncle Henri lit his pipe to smoke with his coffee. I persisted. "You don't even know what it's like to be a surgeon," he declared between two puffs. "You're not going to undertake a profession you know nothing about! If you sit in on an operation, I'll bet you a bottle of cognac that you don't stick it out." And it was agreed that Uncle Henri would ask his friend C. to let me observe a morning session in the operating theater.

On the appointed day, I presented myself to C. at the Cochin hospital. They dressed me as a surgeon. They put me in a corner where I was forbidden to move or to speak. From there, I watched an operation on a woman's belly. Then one on a man's belly. Then on the shoulder of another man. Except perhaps for a brief shudder at the first cut of the surgeon's knife on the first belly, I did not suffer the weak moment that medical students are said often to feel at their first operation. Far from being unable to stand the sight of blood and raw flesh, I found myself plunged into a world unknown to me, a world of which I was instantly enamored. Everything about it fascinated me: the vigilance, the seriousness of the men and women dressed and masked in white; the authority of the surgeon, whose every order and slightest desire everyone rushed to satisfy; the economy of speech; the precision and efficiency of gesture; the meticulous ritual of the scene; the religious aspect of the silence; the expanse of flesh upon which, in the light of an electric sun, were concentrated the eyes above the masks; the gloved hands, agile, expert, that cut, pinched, tied in an instant; the sporting side, the race against time and against disease. And also a sort of metaphysical revelation: the human body that ceased to be a self-contained object, a vessel sealed and all but sacred in which the mystery of life unfolds; this body could be opened, resewn, cut here, removed there, parts of it repaired, certain of its movements disassembled. What intensity in the operating room! What a factory of hope! I left completely bowled over.

Some weeks later, I took the *bachot* exam in mathematics, then the one in philosophy. The next day I registered for the first year of medical school.

Fear. Fear and hatred. Starting in 1936, I began to feel them oozing here and there. Hatred of the other with, implicitly, fear of the other. The other beyond the frontiers, behind the walls of his factory, with the color of his skin, the shape of his nose, the label of his God. Hatred of the Popular Front, of socialism, of the Jews, of Léon Blum. Explosions of hatred and fear with strikes, occupations of factories; with the attack of the Italian fascists on Ethiopia; and, above all, in Spain, that laboratory of violence, hate, and fear. For a long time the Spanish Civil War had for me an abstract character: the struggle between Good and Evil; the aggression of the fascist and Nazi hordes against a people who had, at last, attempted to emerge from the Middle Ages; a world of fire and blood, the world of Goya that persisted: a sort of black sun that rose only at midnight. I saw in this war the possibility of reconciling what had too often seemed irreconcilable: the old patriotism of my family and the ideas of the Left; the country's interests and the defense of the weak; politics and morality. That is the reason I did not understand the refusal of our socialist government to intervene in the face of this fascist barbarity. Nonintervention confirmed Isi's and Karl-Heinz's arguments: it was the sign of a new rift in what had seemed as solid as rock, the very foundation on which rested France and the Republic. For the first time, I had a sense of our having become a tired people, without force, without soul, not knowing clearly what to believe in, not even looking for anything to believe in. A disenchanted people, ready to accept indignities with no other reaction than idle gossip, than discussion at the corner café.

At the 1937 World's Fair in Paris, the war in Spain took on a new reality with the erection opposite each other of the two monstrous pavilions, that of Nazi Germany and that of Soviet Russia, each calling for a socialism. On one side, barbarity trying to be pure and hard, violence naked and openly proclaimed. On the other, a violence just beginning to be suspected in rumors of bizarre trials, the execution of marshals. Two truths, each so sure of itself that it sought to enslave anyone who resisted it in any way. Two ideologies so fascinated by mysticism that each sought to kill the other, despising with the same scorn everything that seemed soft and weak. And then, farther along, the pavilion of Spain, of the Spanish Republic, of which I recall only one feature: the great black and white panel of Picasso's *Guernica*; a convulsion of men and beasts under fascist bombs.

An August Sunday in Etretat. After a swim, I take Odile and a girlfriend of hers for a glass of tomato juice in the casino bar. A cloudless day, the sun sparkling on a calm sea without a ripple. The crowd of vacationers, the happy crowd, in the streets, on the beach, on the sea wall. Near us, a great table under an umbrella. People are arguing after mass. White dresses. Light-colored suits. Ties. Hats. Odile, her friend, and I are talking about the theater: Giraudoux's *Tiger at the Gates*; Louis Jouvet, whose voice I never tire of imitating.* From the next table come isolated words: "priest . . . sermon . . . not energetic enough . . . the Virgin . . . Sister Angélique." Odile's friend is telling us about her plans, her hope of going on the stage. Suddenly, a jovial interruption: old S., bulldog in hand, who annoys my father so much. He is exultant, old S.; he rejoices as he rejoins his friends at the next table. He has just had a call from Paris. Franco's troops have entered Santander. The vise is tightening. The end is near. At the table, then, there is exaltation. In a burst of emotion, the men rise, the women applaud. They congratulate one another. Hugs all around. Champagne is ordered. Corks pop. The wine fizzes in the wineglasses, and Franco's victory is toasted several times over. The three of us, Odile, her friend, and I, are silent, all choked up. Around us on the terrace, the strollers have halted. They look at the group who are drinking and toasting. They wonder. And old

---

* *Tiger at the Gates* (*La Guerre de Troyes n'aura pas lieu* in French) is Jean Giraudoux's play about the Trojan War and the problem of war and peace. Louis Jouvet was the most prominent theater director and actor in prewar France.

S. continues to hold forth, raising his voice for everyone to hear. He comments. He describes the courage and the superiority of the German troops, particularly their aviation, a certain dive bomber who did wonders. That was necessary. That was necessary to put an end to the *Frente popular*, to this communist and anarchist scum who disembowel choirboys, rape nuns, burn churches. All this Bolshevist rot unleashed. For the Russians intervene at least as much as the Italians and the Germans. So that means the German army is superior to the Russian army. One thought as much, let it be noted, but confirmation doesn't do any harm.

And, with all this, the sun does not grow dim. Despite the corpses and the bombs, it is still a day of light, a day of celebration. On the causeway, the crowd stretches and laughs without a care. Nevertheless, Franco's victory is the victory of death.

The unbearable odor you could never shake off. The formaldehyde that grabbed you by the throat, that stung your eyes as you entered the long white room. The two rows of lead-topped tables with a cadaver laid out on each. Around them stood the first-year medical students, each dissecting a limb. Nothing was wasted. The same cadaver was used also by the second-year students, for whom were reserved the head and the trunk. The important thing was to be assigned a thin body, dissection being difficult when everything was engulfed in fat. It was thus necessary to avoid cadavers that were too plump, most of them women's. If you gave a tip to the attendant on the first day, you would have no difficulty getting a lean and scrawny "subject."

I was at a table with two other men and one woman. One of the men never managed to get used to the atmosphere, the smell, the cadavers, the endless commotion of a hundred students buzzing about, discussing. One day he nearly vomited into the viscera of the gutted body. Our woman friend, on the other hand, adapted perfectly to the situation. Cutting, digging into an arm didn't stop her from leaving as fresh as when she walked in. Fate had given me a leg. And, after a bite of lunch eaten in haste with friends in a little restaurant in the rue Racine, I found myself facing "my" leg, which I was to dissect starting with the upper part, the groin. First, I had to make an incision the length of the thigh. Then, to clean the subcutaneous space to lay bare the first layer of muscles. To resect certain of these muscles to expose the femoral triangle. To locate the femoral nerve, then the femoral artery and veins. To isolate carefully these latter so as not to break them. Slow work demanding patience and dexterity; training the hands to be delicate and meticulous: a first step toward the surgical profession.

But also, a terrible apprenticeship of life and death. Of life through death. For the first lesson to be drawn from the anatomy pavilion was that, in order to do medicine, to take care of the human body, one had to know this body; and that to know it, one had first to destroy it. To destroy it wittingly, calmly, methodically, step by step, organ by organ, while always digging further. To manage this, however, to work in this anatomy room, one had to forget what, in these bodies, had been human. Forget that, but a short while ago, they were walking and talking. Not wonder what they had accomplished or dreamed of accomplishing, why they had ended up this way, in this room, on one of these lead-topped tables. But if it was true that man defines himself by his knowledge of his own end, that he distinguishes himself by the tribute and respect he renders the dead of his tribe, how then tolerate the treatment inflicted on these cadavers in the anonymity of the anatomy pavilion? How to accept these bodies transformed into an object of study; these tortured, slashed limbs reduced to shreds; these organs cleaned, stripped of all connection; these structures whose rigidity obscured their original function; this calculated destruction of what was the flesh of a man or a woman. And, then, what were the backgrounds of all these poor old men, these poor old women? How many among them had been rich merchants, politicians, high civil servants? What did they do to end up here? What crimes had they committed to deserve such punishment? Had they "donated their bodies to science," or had science become a garbage receptacle for those whom society was no longer interested in, who no longer produced or even consumed? What order of the world justified such contempt? And, for this human debris, what Antigone would ever come to demand a decent burial? Still, however weighty these unanswerable questions, however trying the repeated engagements with these nameless cadavers, anxiety and reluctance gradually petered out. Interest and curiosity soon won out over the unbearable sense of human destruction. And after the afternoon's dissection, one wrapped the "stiff" in sheets that had been carefully dampened to keep it from drying out.

With the transformation of the lycée pupil into a medical student, a new life began. New because of the sudden freedom. Because of having a status more closely connected with life in the real world. As much as the lycée cloistered its pupils, the medical school gave its students their heads. They did what they wanted. They could just as well live in the cafés of the Latin Quarter as work night and day. For my part, I considered the surgeon's profession the finest, the most noble in the world. And I had no doubt that to become a great surgeon meant a lot of work. Quickly enough, I discovered the empirical nature of medicine. The human body had its strength and its beauty. It also had surprising weaknesses, the reasons for which were obscure. So it was. One had to understand it. But, except for some infectious diseases where the microbe responsible was known, though not how to combat it, there was little apparent logic to illnesses. Their cause was impenetrable. Descriptions had to be learned unthinkingly, by heart, without resort to reason. This activity culminated in studying for the examination for hospital outpatient clinic. The rank of *externe* was the lowest in the hospital hierarchy. To make it, one had to take a competitive exam in the third year of medical studies, in which one had ninety minutes to write out the answers to fifteen questions: on medicine, surgery, anatomy. Six minutes a question. No time to reflect. Just spit out the answers as they came. To prepare for this exam, little groups of students gathered in "seminars" under the direction of an intern. We spent most of our evenings this way, identifying questions, memorizing them, regurgitating them on demand. Medicine in spare parts.

In my memory, these first years are indissociably linked with the laboratory of my anatomy professor, Monsieur Hovelacque. Each year he took on a dozen beginning students to whom he gave a particular kind of instruction. Most of these students were the sons and daughters of his colleagues, leading physicians. I had access to this in-group thanks to my grades on the PCB* examination, the best in my year. A horde of people swarmed around this laboratory, jammed with cadavers, skeletons, bones, medicine bottles, books. There were students like us seeking to learn the rudiments of anatomy; young graduates in search of new operating procedures; experienced surgeons intent on improving their techniques; candidates for the surgical examinations, for the posts of assistant and of demonstrator, ever desirous of showing off their knowledge, of rehearsing for an oral exam, of criticizing each other. And over all this world reigned the master, benevolently but firmly. Very tall, very thin, with an emaciated face above a long white beard which he continually stroked, eyes glinting with mischief from behind steel-rimmed glasses, always wearing a white lab coat and apron, Monsieur Hovelacque had the majesty of an El Greco cardinal. This ascetic spent some twelve to fifteen hours a day in the laboratory. It was said that, as a doctor to the Algerian Zouaves in the First World War, he had, when the war was over, returned directly to the laboratory before going to visit Madame Hovelacque whom he had not seen for months. Seated on a table, rolling his own cigarettes, he loved to have a tête-à-tête with a student, to ask him about his plans, to give advice. Thus, one day he made a suggestion to me: "Jacob, if you want to be a surgeon, don't get married. No little woman or you'll have had it! A little sex here and there, but that's all!"

At the time Monsieur Hovelacque's main activity was preparing a surgical atlas of the anatomy of the thorax and, to this end,

* Physics, chemistry, biology, the first year of medical studies.

he was making highly detailed analyses of horizontal cross-sections of the thorax of a man who had been recently guillotined. With these cross-sections, it was easy to specify the best way to get at a particular organ. Monsieur Hovelacque scarcely dissociated his teaching from his studies, having an equal passion for each. Though he drew poorly, he had devised a system of proportions so well developed, a game of reference points so precise, that he could, in a few minutes, draw any region of the human body on the blackboard. His great specialty was osteology, the science of bones, about which he had written several books. For him, a bone as simple in appearance as a clavicle became a fantastic landscape whose mountains and valleys could be traversed ad infinitum; whose every spur, every groove, every muscle's point of attachment, every relationship to other organs could be described down to the last detail. His course was a lesson in simplicity and precision; the epitome of the art of description. More than a few of the students were afraid of Monsieur Hovelacque. They took him for an odd fellow, persnickety of mind and particularly vicious in exams. But they knew only his exterior, the imposing, disconcerting side of him. The man harbored a passion to which he devoted himself body and soul. But even someone with only a moderate taste for anatomy could not but admire this austere, solitary life, wholly bent on disinterested study, on the pursuit of knowledge. It was in him that I first saw the grandeur and the servitude of research.

The most fascinating and the most dreaded part of the first year was the hospital training. The apprenticeship that now could be done only on human beings, on living people. The entrance into the real world of medicine with its wards, its iron cots, its smells, its nurses, its hierarchy of doctors who, at every rung of the ladder, called their superiors *Monsieur* and their inferiors *Vieux*. The first contact with patients. The shock in the face of suffering, as brutally discovered as was our inadequacy to remedy it. The eruption of misery, of a distress as much physical as spiritual, which appeared suddenly in a glaring light, in the nakedness of a body one uncovered to examine. The terrible nakedness of these men and women, anxious, embarrassed, ashamed of being there, the body ill, emaciated, blighted.

In the hospital wards, the fifty first-year medical students were like a horde, a bleating flock who got in everyone's way and were confined to certain limited areas. Each "trainee" was assigned one or two beds. Upon the arrival of a new patient, the student took the "observation," recorded his medical history, his childhood illnesses, the health of his parents, his style of life, the amount of wine and liquor he consumed, the presenting complaints. Unbearable incursions into private lives. Peeps through the keyhole. There were also presentations of the patients: the crowd of doctors, nurses, medical students around a patient, isolated, lost, given over to the eyes of others, kneaded, manhandled for hours, while the chief physician or the head of clinical services droned on interminably. And then my first patient! A little old woman with a racking cough: wrinkled, stunted, yellow, and scared, her breasts flaccid and pendulous, her hair thinning. I stood there before her, not daring to touch her, not knowing how to auscultate with my brand-new stethoscope, awkward in my gestures under the withering eye of the nurse. The old woman had pneumonia. She was dead before the week was out.

For want of a definite role, we sometimes had to take risks. In the first year, the student was caught between two attitudes: on the one hand, the diffidence imposed by youth and ignorance; on the other, the daring that comes from a desire to learn. So we lived in an ill-defined space: a tense zone between dream and reality.

On my first day of surgical rounds with Professor G. at the Saint-Antoine hospital, I meet an *externe* who immediately leads me into the operating room: "*Vieux*, you are going to do the anesthesia." I find myself squatting behind the head of a female patient who has been tipped back to expose the pelvis where the operation is to be carried out. The *externe* shows me how to place the mask which gives off a strong odor of chloroform and ether. Turning the lever one way increases the dose of anesthetic; the other way decreases it. To gauge the depth of anesthesia, watch the pupil of the eye. Push the jaw forward so the patient doesn't swallow her tongue. Note the regularity of the respiration. The air bladder that swells up and empties out in rhythm. A sudden growl of discontent from the surgeon. "Your patient is not sound enough asleep." I hastily increase the dose. The pupil. The air bladder. Another growl from the surgeon. I increase still further. Impression that the patient is breathing less well. That the intake of air is less regular. That the air bladder is inflating less. I decrease the dose a little. The pupil dilates, dilates. Anxiety grips me. The air bladder is almost still. I clutch at the mask desperately. The surgeon and his assistants are looking at me. The *externe*, who has gone off, comes back. He shoves me aside and takes my place: "The patient has gone into syncope!" He grasps her tongue with the tongue-holder and pulls. The breathing returns, uneven at first. Then full, then normal. At last! What fear!

Munich and the abandonment of Czechoslovakia. The die was cast. Then began the unbearable wait. Peace on its deathbed. A world ending, the warm, sweet world of my childhood. Then also began my mother's illness, during Munich, or shortly afterward. What was still only a little lump in the groin became a malignant tumor the day when, arriving at my Uncle Henri's in answer to his summons, I found him looking somber, frowning. He had wanted to see me before speaking to my parents. He looked at me for a long time. He must have been wondering whether he still had to treat me like a kid and tell me stories; or whether now like a colleague to evaluate the situation. Finally, he took a sheet of paper from a table and handed it to me. The diagnosis was bad: a form of cancer of the lymphatic system. It was going to be necessary to operate to remove all the suspect tissue. Later, there would be X rays to eliminate the last cells. A great deal of progress had been made in this area, he assured me. He would refer my mother to Professor F., the leading specialist, a very gentle, very understanding man. There were cures now. We had to cling to hope. Hope. Hope. For several long minutes, we stayed there without speaking, without daring even to look at each other.

I wandered through the streets. Often, after receiving some blow, one's pain holds off for a time before bursting out. In the same way, after the abrupt news of the death that was lying in wait for my mother, I was in shock, without yet feeling its effects. I sensed an immense grief was about to explode in me. But I did not yet know that I would have to get used to seeing my mother ill, to thinking of her as lost. I did not yet tell myself that soon I would never see her again; that I would have to help her overcome her anguish, do what I could to make her suffering less cruel. I walked for a long time. Night had fallen. It had rained. The sidewalks were still wet, but the moon now shone in a cloudless sky. I tried not to think. There were a few cars. Only a few sounds. The trees were beginning to shed their leaves. Out of the darkness loomed a girl. She accosted me and took a few steps toward me: a young face with apple cheeks; a sensual smile showing wide lustrous teeth; small, high breasts. She stopped at the door of a hotel. "You look sad, come up and tell me about it." Drown my anguish in her? I nearly went in with her. But I was afraid of the aftermath: my disgust with myself. I went on walking, alone.

When I got back to the house, my father was already in bed. Sitting in the living room, her needlework on her knees, my mother seemed absorbed in a kind of reverie. For an instant, I could gaze on her while she was still unaware of my presence. She who was always so cheerful, so animated. I now found her thinner, with a mask of fatigue and lassitude that I did not recognize. Perhaps the news of her illness made me look at her in a new light. Abruptly, she had what I had never yet known in her: age. As soon as she saw me, her face was transformed. In a flash she looked younger, all lit up with smiles, delighted to see me home early. Poor maman. I wanted only one thing: to rush into her arms. To cover with kisses that face sculpted in tenderness. To keep it as it was, living, palpable, real, forever. At the same time, I felt tears welling up, which I didn't want her to see at any cost. Brusquely, I snapped: "I've got a lot of work to do. See you in the morning." And I left the room.

Inexorable, the march to war. Inexorable, the illness of my mother. Both of them forever linked in my memory. Munich and the discovery of the cancer. The fall of Barcelona and the operation. The occupation of Prague and the X-ray sessions. The Phony War and a temporary remission. Knowing I would soon not see maman again, I made an effort to get home early to find her, to have her to myself, to humor her a little. In the evening I often went to sit on her bed, to tell her about my day. I felt I was providing her with a little of the tenderness she had lavished on me as a child. She sometimes seemed to avoid me. She shut herself up with my father or my grandmother or one of my aunts for long secret councils I was forbidden to interrupt. I would be a bit put out with her for not greeting me on my return. It took me a while to realize that she refused to see me only when her pain was too great and she felt ill. That she also waited for me impatiently, but preferred to do without me rather than leave me with a bad memory. She died in early June 1940, at the height of the German attack.

17 June 1940. One never forgets the day of one's twentieth birthday! I am sitting in a low car, a black Citroën 11, with three of my contemporaries, comrades from the lycée or medical school. The car is going along a road in the center of France, in a southwesterly direction. We have been taking detours to avoid, so far as possible, the vast snarled mass of refugees, a river of torment and fear that moves slowly south. Despite the early hour, the weather is already bright and warm: one of those days filled with light, a festival of nature, so propitious for the maneuvers of airplanes and tanks. The car is going through the countryside on a narrow local road which is nearly deserted. Overcome with fatigue and disaster, we keep silent, each attending to the disorder of his own thought. Still too young to be called up, I

have finished, after a fashion, a second year of medical school, distracted by the war and my mother's illness. A week ago, right in the middle of exams, we learned of the Germans' imminent arrival in Paris. Now, while they were entering by one of the Paris gates, I was leaving by another to rejoin my father, who had taken refuge in Vichy with my two grandmothers. Early this morning I had left him, overwhelmed by the avalanche of misfortunes. Poor Papa, I can still see his silhouette against the white wall behind him, while on the verge of tears he slowly waves goodbye.

The car has rejoined the national road. It goes slowly, ensnared in the throng of trucks, cars, carts, bicycles. In the sun and dust, the road is teeming as far as the eye can see. From time to time, a military truck passes the column, honking its horn. What an incredible mess! In a few days, I have seen a whole nation disintegrate. For twenty years, it has been as though I were learning to play a game. Constructing a world around myself. Erecting piece by piece, bit by bit, first the milieu around me, the world of the everyday, with my room and the house, the city, the lycée, the university, their pasts and their histories. Then understanding the country, the Republic with its institutions and its laws, its army and its justice. And suddenly the whole edifice has caved in. Everything I believed in, everything that I thought I'd believe for life, everything that seemed the very basis of our existence, forming our protective armature, that seemed to shape our view of the world: all this crumbled in an instant. In an instant, the country has foundered. In an instant, despite its great men and its great schools, its generals and its institutions, its teachers and its senate, it has collapsed, body and soul.

And all this on account of a single madman. This nightmare, because a raving lunatic with haggard eyes, foaming at the mouth and screaming maledictions, had coolly decided to put the world to the test of fire and blood. And because governments of bunglers, manipulated in the shadows by degenerates, did not know how, or did not dare, to stop him. Or even, thinking they were clever, had let him do as he wished; had perhaps encouraged him, imagining that they could make use of the rabid monster, let him destroy what annoyed them, later to capture him and bring him back to heel. Idiots or criminals? Fools or knaves? In 1935, Hitler could have been stopped dead in his tracks. The occupation of the Rhineland was a matter of poker, a bluff. One word would have been enough. No. *Non. Nein. Nyet.* And the wild beast would have returned to his lair. One would no longer talk about that. But that was too simple. That did not suit our society of fastidious, sheltered citizens; nor our governments of profiteers. They were obsessed by the man with the knife between his teeth. The Bolshevist bear. There lay the danger: the red beast at the door, invading our factories, sowing the seeds of the Popular Front. And, to destroy this red beast, let's throw the black beast between its paws: they will destroy each other, and the game will be won. That's political genius. Too bad it didn't work. The two savage beasts, red and black, joined forces to devour those who would tame them. And once again, it is the poor types who will pay: the lamplighters, the workers, the Jews. As for the sons of bitches, they'll get away with it, as usual.

Still the same heavy silence in the car. Outside passes a stream of fields, cows, forests. A country of dream. A country of plenty. Handed over, neatly wrapped, to the Nazis. What a crock of shit! The car threads its way as well as it can among the columns of fugitives, the trucks, the carts. Sometimes, high overhead, a plane surveys the scene. In the car, Roger every now and then whistles a few notes from the Charles Trenet song *"Y a d'la joie."* A bit annoying. We stop for lunch in a village in the Auvergne. In front of a little inn, the four of us flop down at a table. Suddenly, from an open window, we hear a quavery voice, the voice of Marshal Pétain, charged that very morning with forming a new government: "Frenchmen, it is with a heavy heart that I say to you today that the fighting must cease. . . . I have asked for an armistice with honor and dignity."

When the car takes off, a great deal of excitement reigns. Behind the anguish, the voice of Pétain has liberated passion. The first, Roger, is blazing. Roger D. is an old friend. We used to play marbles in the Parc Monceau. For a long time, we were in the same class at the Lycée Carnot. I had a lot of admiration for him, for his self-possession, his lucidity, his humor. Also for his athletic ability: he played on a rugby team. Finally, for his good fortune: attractive girls often awaited him at the lycée door. "My big brother sends them over," he said modestly. As soon as we're back in the car, Roger lashes out: at Pétain, at the military, at all the incompetents who had fallen down on their job of preparing for this war. Like the traitors, the crooks, the dirty bastards of every stripe who want a Hitler in Paris to shape up this country and re-establish an order threatened by the Popular Front. He warms to his theme. He raises his voice. He hears himself shouting. All those swine. Those grafters. Blind men. They were told what would happen. It was even written down by Hitler himself. He wrote that he would crush anything that got in his way, that he would grind his enemies to a pulp. That he would make them slaves of the Reich. But the French didn't give a damn. All they cared about was the cushy little job, the apéritif before lunch, the little wife on Saturday night, vacation, a game of *pétanque*. Even so, we're not going to let ourselves be had. We're not going to wait here for the SS to arrive so we can smile prettily for them. When they are here, that will be the end. That doddering old fart Pétain won't keep them from doing what they want. You don't talk it over with the Nazis. You bash their faces in. There's only one thing to do: go on with the fight. And to fight, you have to get out of France. Go where you can.

I am in total agreement with Roger. I can only support him. I also produce my tirade, but less violently. Less clearly, too. More fatuously. My arguments are mainly military, no doubt because of my upbringing, because of the military tradition throughout generation after generation of my family. Because of the influence of my grandfather the general. Because of the story of *le Grand* in 1870. Because of my father and my uncles, each telling about his battles. Today, then, it is our turn to play. All the more as the military threat is based on an intolerable political system of domination and repression. But we are not going to shrink before a threat worse than any ever seen. Worse than Genghis Khan. Worse than Attila. You don't negotiate with a Hitler. Either you destroy him, or he'll destroy you. I, too, am for getting out of France, for fighting wherever we can.

The two others, Michel C. and Maurice B., hardly react. They are rather embarrassed, uneasy. Michel is surely not a hero. Too large a head for his short legs. Cunning. Clever with his fingers. He, too, wants to be a surgeon. Still something of a schoolboy, he loves song, wine, easy girls. Has little interest in politics. His family was rather reactionary. What he hears in the car, particularly Roger's imprecations, does not convince him. That's all so much gibberish, claptrap. He's not going to yield to passion. There's time to see what will happen. Say what you will about Pétain, he's still a guarantee. As for North Africa and England, how can you make plans? Who can say what's going to happen there? Michel looks to Maurice for support. In vain. Maurice doesn't open his mouth, but stares at the horizon through the window. Half little boy, half parish priest, Maurice loved equations and the gentler sports. He had a collection of crystal and carefully maintained his virginity. Skillful with his hands, he was the only one among us capable of repairing anything: a lamp, an alarm clock, a car. A loyal and faithful friend in good times, he proves disappointing in bad ones. That day, as in the ones to follow, his behavior is selfish and cowardly. In the muddle of catastrophes, he always keeps his sights on getting into the Polytechnique. What interests him above all is what his papa will think of the situation. We constantly have to stop for him to take a piss by the side of the road.

At the end of the day, the car reaches the resort town of Arcachon on the Atlantic coast, near Bordeaux. There I find Uncle Henri, who has arrived two days earlier with his wife and daughter. What an impression of stability, of solidity I feel in Uncle Henri! What a comfort, amid the shambles, to find this giant of a figure with his grizzled beard, his gigantic ears, the warmth emanating from him, from his eyes, from the expression behind his glasses. What I have come to get from Uncle Henri this evening of 17 June 1940 in Arcachon is, before I leave France, the blessing of one of my elders. We are walking side by side on the beach. Night has fallen, a night of blackout when only the stars are twinkling. In the lingering warmth, Uncle Henri, hat pushed back on his head, has undone his vest. He frequently scratches the hairs on his neck, under the beard that he combs with spread fingers. His grave voice, with a smoker's hoarseness, dominates the lapping of the waves. As Herriot's doctor, he had leisure to witness the many months of intrigues and plotting in the government. He tells of the weakness of political men, Pétain's double-dealing, General Weygand's trickery. The struggle over whether to continue the war in North Africa. The atmosphere of treason prevailing over Bordeaux with the coming of the government. The imminent arrival of German armored columns, which are advancing down the Atlantic coast at full speed. He stops suddenly and looks me in the face. He is offering me not just his blessing but the full weight of his authority, the support of all his clear-sightedness. He puts his hand on my shoulder: "If you can, go to England rather than Africa."

Early the next morning, the four of us get back on the road. Roger and I want at all costs to find a boat leaving. First in Bordeaux, now an immense caravansary with the arrival of the government and the civil service. Then on to Bayonne. On the door of the British consulate, which is closed, we find a note: "For all mlitary inquiries concerning the French, apply to Saint-Jean-de-Luz, such and such a street, such and such a number." Disheartened, fearful, Maurice leaves us there. He wants to find his papa. Wavering, Michel accompanies Roger and me to Saint-Jean-de-Luz. In the street, specified on the outskirts of town, there appears to be nothing of interest. We wander around. Still nothing. At the end of the street appears a cavalry lieutenant in a blue képi, who comes up to us and asks: "Are you looking for something?" "We've been told our cousin Marie is staying here," answers a suspicious Roger. The officer smiles: "You want to go to England?" No response on our part, but a certain interest. "So," the lieutenant continues, "come to the port around five this afternoon. There are two ships at anchor. Polish troops are going to be embarking all day. This evening, when they have finished, it will probably be possible for you to leave with them. Don't make yourselves too conspicuous before then. Good luck." And he vanishes. For several hours, we wander around, a little feverish; first in the back country, then in the little town. Michel finally decides to embark with us. But not for long. As we approach the port, he happens to run into an uncle of his. After a brief talk, the uncle forbids Michel to go. "You're not going to do anything so silly, running away with Freemasons and Jews." Which Michel has no burning desire to do, anyway. Exit Michel. We stay. Just the two of us, Roger and I.

Five o'clock. The little port of Saint-Jean-de-Luz is lit by the sun of a summer that begins this twenty-first of June 1940. On the quay, a crowd of military men and civilians is waiting. All day long, fishing boats have been ferrying the remains of two Polish divisions, who had fought on our side, to board the two ships anchored in the harbor. Roger and I fall in with the harassed, tense civilians hoping to find a way to get out of France. Some faces are resolute; most of them, anxious. A crowd surprisingly calm, even passive. Nothing disturbs the silence but the breaking of waves, the sound of steps, and the motors of boats. As the last of the Polish soldiers embark, a cordon of gendarmes position themselves to prevent the French from departing. A slow shuffle to reach the end of the quay and board the fishing boats. In front of us is a short man, a jockey, I later learn. With his laughing eye, his obvious disguise as a civilian, his lower-class Parisian background is plain at a distance of three hundred feet. A towering gendarme sees him coming and puts his hand on his shoulder: "Just where do you think you're going?" The man bristles, looks straight at the gendarme, and yells: "*Svastika,*" the first Polish-sounding word that comes to mind. Dumbfounded, the gendarme lets him pass. Roger and I take advantage of his surprise to sneak onto the boat. A half-hour later, we board the Polish ship S. S. *Batory.*

That evening, we find ourselves seated side by side on the deck with the little jockey and another passenger. England? North Africa? No one knows where the ship is headed. "Ever hear of de Gaulle?" the little jockey asks suddenly. And, without waiting for an answer, goes on: "He's a general. I heard him on the radio. He said he is going on with the war, in England. He said that, sooner or later, we will beat them. Others bow down before Hitler. So things are simple, no?"

On this first night of summer, the stars come out, indifferent to human history. Not a wave on the sea. Not a breath of air. Only the drone of the screws breaks the silence as we slip away from a coast where, in the dark, the German divisions advance. I have no idea when I will see this coast again. Or even whether I ever will.

IV

THAT dark line—still faint, still blurred in a blue haze, far away on the horizon—is the coast of France. The line is still unreal, like this instant, so long awaited, so long dreamed of, so long cherished. A moment of suffocation in which laughter mingles with tears, when events experienced suddenly reproduce the event imagined. When what long seemed impossible suddenly materializes to overwhelm us beyond the possible, denying in the instant four years of exile, anguish, solitude, fighting, despair. Four years that are converging on this dark, but already more distinct, band on the horizon, becoming more precise, expanding slowly with the rhythm of the boat. The soil of France! The Promised Land. Cockaigne. Paradise. Eldorado. Just like this little island that was described in detail in a picture book from my childhood, with its countryside so welcoming, its climate so benign, its flowers so fragrant, its fruits so succulent, its inhabitants so friendly and so loving that no one could wish to live anywhere else. Before us, less than a mile away now, lies the land of my childhood. The incredible in its pure state. The marvelous in the raw, absolute, sufficient to take your breath away. The return home after four years, as in a reverie whose drift one passively follows. The pariahs suddenly become heroes, dragon slayers. Beyond the beach, sand dunes appear more clearly outlined in the sparkling light of this 1 August 1944. A summer as vivid, a sun as sparkling, as in June 1940. The same sky of fire and glory. This time, however, it is our planes flying and the enemy planes that are falling. This time, however, the men I left with will not be returning: not Roger D., killed in Chad from a spear in the heart; not the little jockey, killed at Bir Hacheim, in Libya.

In Southampton the evening before, our medical company, one of the last units of the French Second Armored Division (known by its French initials, DB) to leave England, had boarded an LTC, the flat-bottomed landing craft. A eventful sailing. Zig-zagging in the harbor, we twice managed to butt up against two other boats moored in the roads. Drunk, the British captain commanding the LTC, a great strapping, red-bearded fellow in a white uniform, lurched about on the bridge. In the light of the setting sun, we finally managed to rejoin that veritable boulevard of the sea along which, night and day, thousands of ships incessantly coming and going under the protection of innumerable guard dogs, torpedo boats, and airplanes linked England to the Normandy bridgehead. In the middle of the night, half asleep, we were wakened by a sudden silence. No more motor. No more drone of the screws. Only the waves slapping on the hull. The sun rose on an empty sea. Not a single boat on the horizon. Not a plane in the sky. No one on the bridge. Our LTC gently floated with the swell. All we could do was wait, hoping no coast guard, no enemy plane would have the bad taste to arrive in the vicinity. Two hours later, the captain appeared on the deck. He stretched luxuriously; downed a can of soup for breakfast; and then, armed with a telescope, slowly scanned the horizon. The motors started up. A short time later, the cohort of boats and planes emerged from the fog, swarming in the distance. Then the coastline appeared.

What a homecoming! Beyond anything we had dared imagine for four years. My eyes are still filled with images. Images of sand and sun. The sand of this little beach which the Americans had transformed into a great industrial port. The sun of those heavy August days that had become, for us, so light. And then the villages: Sainte-Mère-Eglise and its steeple on which the invading D-day parachutists had gotten caught. Normandy and the war. The farms surrounded by hawthorn hedges and the wrecks of incinerated trucks. The narrow tree-lined roads and the white-helmeted military police. The smell of hay and the sound of cannon in the distance. The mornings dawning as clear and as pure as though there were no war, and the fleets of planes taking off to bomb Germany. The gnarled apple trees and the columns of German prisoners of war. The glass of cider on a farm and the American K-rations. The distrust of the country people and the ardor of the men of the DB. And the countryside: all light, all smiles despite the buildings gutted by bombs. And the fear of mines. And the rumors of isolated snipers hidden in trees. And my driver's constant whistling of "Lili Marlene." And, after the storm of iron and fire, the harmony that seemed to be reborn between men and the earth.

Frozen images. Dead images. In vain have I tried to summon up sounds, smells. I cannot recapture the intoxication, the giddiness of those days. I recall landscapes, faces, scenes. But these memories remain flat, so to speak. They lack depth. They have lost all the sap that nourished this period and that, for those who lived it, gave it its unique character. How can I reproduce today the intoxication of that return? How to retrieve the state of exaltation in which, along with the sweetness of treading anew a ground long thought forever forbidden, mingled the violence impelling us to fight? How to reconcile the rush of feeling that came with suddenly understanding again in every limb, in every muscle the words "my native land" with the passionate desire to cleanse this earth of all the mornings of agony, all the nights of prison, all the days of anguish and humiliation suffered for four years. How to revive that state of triumph and vengeance, of joy and fury? How to experience today the sense of indestructible cohesion we felt in the DB, attached to the American Third Army, Patton's army? Or the irresistible high spirits of the roughneck soldier, the feeling of brute force which nothing can resist? How to feel again the conviction, then rooted deep in each of us, of witnessing the destruction of one rotten world and of arriving in time for the reconstruction of another, new and beautiful and generous? In short, how can I re-create now the certainty we had that we were making history? Or, better yet, that we were actually *being* history, in the raw!

For me, however, the euphoria of the return did not last very long. Only a week. A week in which our medical company went from field to field, first near Sainte-Marie-Eglise, then in the Mortain area, refurbishing our equipment, checking out the cars and ambulances hidden among the trees. A little boredom. Much impatience and enthusiasm. As if this pre-combat vigil had suddenly induced a new mood, in which gaiety joined hands with gravity. Many of the men had not yet been under fire. Some were nervous. But all of them dreamed of seeing action. Which prompted Lieutenant B. to irony: "I suppose you want to smother them in bandages! Bash in their heads with stretchers!" Tall, dark, with an emaciated face, black eyes magnified by glasses, B. had a dry sense of humor which was not to everyone's taste. We had become good friends one day in England, before leaving for Normandy, when we were surprised to find ourselves on the doorstep of the same young woman, each unaware of the favors she was giving the other. Fate later assigned the lady exclusively to B.

On 8 August, the order finally came for the company to move out toward Le Mans. Patton's army executed a vast maneuver to take the German troops in Normandy and Brittany from behind and so open the way to Paris. We had to wait for nightfall before getting under way. All day long we heard gunfire intensifying to the east. The Germans were counterattacking in the direction of Avranches and Mont-Saint-Michel, in an attempt to cut off the American Third Army in its advance to the south. That evening, after the camouflage was removed, the vehicles started forming a column. The sergeant-nurse who was driving my truck was furiously whistling "Lili Marlene," as though to announce to the world that the Second Medical Company was getting under way. Suddenly, before the column had even started moving, the characteristic droning of the German bombers was heard in the distance. The sound grew louder. The planes were approaching, invisible in the night. When the first bombs exploded in the nearby field, the sergeant and I just had time to jump from our truck and throw ourselves into a roadside ditch. A series of whistling sounds transformed into thunder. Flashes of light. Shouts in the night. Explosions that shook the field. Brief flare-ups on cars, one of which caught fire. In that night of madness, I lay motionless in my ditch. Glued to the ground. Hugging it close. Not budging. Not letting an inch of myself show above the ditch. Then the convulsion stopped. Silence returned, a silence soon broken by the cries of the wounded.

Several men remained on the ground. Not far away, lying near to a car, his features contorted, Lieutenant B. was moaning softly. A bloodstain was spreading over his side. I smiled at him. I wanted to joke. He looked at me sadly, without saying a word. I began to rip open his jacket and shirt to apply a bandage. When we tried to lift him onto a stretcher, he screamed. At that moment we heard again, from afar as before, the regular drone of the returning Junkers. The sound of the motors grew louder. His face straining, B. tried to get up; then, grimacing with pain, he fell back again. The ditch was less than thirty feet away, but B. could not be moved. His eyes wide with anguish, he took my hand: "Don't leave me." Everyone around us was diving for cover. I looked longingly at the ditch. There was no question of dragging B. there or of abandoning him. When the first bombs whistled, I snuggled against him, as much to make myself as small as possible as to shelter him a little. We remained there tucked up against each other, holding our breath. Motionless. For the second time, the earth shook, heaving in a clap of thunder. With clods of earth flying in every direction. So that I found almost normal the violent jolt I suddenly felt on my right side. It was, so to speak, part of the disorder that had come in the night to disturb this little field in Normandy. A violent shock, but no pain. For an instant I stayed there without moving. An instant I tried to make as long as possible, as if knowing already that nothing afterward would be the same. A sticky black spot appeared on my right elbow. I tried to move my hand. It hung inert. I tried to get up. In vain. Only then did pain descend: sudden, massive, intense. I passed out.

When I came to, I was in an ambulance whose every bump unleashed agonizing pain. Later, I regained consciousness under a large tent in an American field hospital. On the next bed lay B., pale, still, more emaciated than usual. When I regained consciousness once more, B. was no longer there. I learned he was dead.

The memories that recur to us are filled with silence. Even if the reality that they evoke is teeming with noises, those have disappeared. The field where I was wounded, the night, the flashes of light, the explosions, B.'s suffering: I remember all this in profound silence. I *know* that I heard the detonations, the great uproar invading the night, but I hear them no more. At best, I can reconstruct a sort of distant, muffled hubbub, rather like films of the silent era to which sound is added today before they are shown. In these memories live only mute apparitions, shapes expressing themselves with gestures and looks, without recourse to speech. Perhaps memory blots out the words and the noise to eliminate the superficiality and chatter, to keep only the kernel of life, to seek to attain what we think are the silent powers hiding behind things. Or perhaps this silence acts, on the contrary, like an infirmity of memory, the images of a past amputated from what animated it, evoking in us sadness rather than desire, the infinite melancholy of memories. Things were. They will not come back as they were. The secret forces, all the murmuring of life that our past arouses in us, cannot be reborn. When we look at the photo of someone who was our friend, what forms in our mind are his features, his face, the days we spent in his company. But it is not he, with his voice, his energy, all the sounds and the secret signs that emanate from a young body. Even if memory were to restore the landscape of our youth to us, it would deliver it silent, dried out. Dead.

Never did I swell my chest out so much or hold my chin so high. Never did I walk so proud or click my heels so smartly on the pavement. My first leave. My first outing as a military man in the streets of London. And especially my first uniform. Battle dress, with a tricolor emblem sewn on my shoulder. In my snugly buttoned uniform, my beret cocked on my head, my eyes fixed on the horizon, my legs straight, I strolled majestically, punctiliously saluting, with elbow held high, every military man I passed, whatever his rank or nationality, as though to tell the whole world of my personal, individual decision to get on with the war. As though my father, my grandfather, my uncles, *le Grand*, the whole line of ancestors who in their time had borne arms, had come, attentive, approving, to watch the way I took my turn. To don this uniform was to reject a defeat that was not mine. It was to give meaning to rage. To transform despair into anger.

It had been a scant two weeks since I had shipped out of Saint-Jean-de-Luz with Roger D. Strange voyage. After the nightmare of the exodus, the nightmare of exile. The Polish ship slipping over an oily sea in radiant sunlight. A smooth crossing, troubled only by submarine alerts. The hundreds of Polish soldiers and the civilians, tired, disheartened, cluttering up the decks and gangways or obediently lining up to fill their mess tins. Dozens of Frenchmen, tense and anxious, attempting to regroup, asking each other questions, arguing far into the night about the treachery or the virtues of Marshal Pétain; about the benefits of the armistice versus those of the struggle carried on in the French territories in North Africa; or about the British ability to hang on or the fate of the prisoners of war. And also, moving from one group to another, there was the inevitable bunch of eccentrics: the man from Toulouse who had cleared out of France "so as not to have to pay taxes to a government of incompetents and losers"; or the one from Bordeaux who had "grabbed at the chance of a lifetime to get away from his wife and mother-in-law."

Then the arrival in Great Britain. The transfer to London in custody, the British properly distrustful of a subversive fifth column. And all that in the English atmosphere where, never since, have I been able to return without emotion: the welcoming charm of the green, green countryside; the rows of little red-brick houses, each one like the next even to the square of close-cropped lawn; the ubiquitous tea with milk; the sweetish aroma of cigarettes. Calm. Assurance. Peace. All the more amazing as we were leaving a nation in total disorder; for our eyes were still filled with images of the collapse, of refugees by the million on the road, of bombardments. What we found on landing was a nation intact, trim, well ordered, scarcely touched by the whirlwind that had already carried off several nations like wisps of straw. Here, the war was muted. Not that England was unmindful of the danger awaiting it. Everyone knew. Everyone expected aerial bombardments, attempts at invasion. But everyone was preparing calmly, confidently. Factories were running around the clock to rearm. Men and women enrolled in the Home Guard. In the eyes of the English, the "Huns," as they called the Germans, had only an insignificant navy. And it was not with airplanes that Britain risked being overrun. One pink and white old gentleman, who had come to the railway station to hand out tea and buns, told us, "We'll be returning your country to you in a few months." While we did not put much credence in these brave words, we were grateful to the English for trying to give us hope.

In London, in a sort of shelter for old people, the transit camp where the English sorted out the horde of people who, by any and every means, sought to flee Nazi Europe. A curious zoo occupied this camp where languages and accents, nationalities and ideologies were mingled. A docile crowd, cooped up. Long lines: waiting for the immigration officials or the police, or in front of the toilets; waiting to get a blanket or a half-pint of tea with milk. And, on top of all this misery, the differences that exploded, often touching, sometimes unbearable. The arrogant and the humble. The well-heeled and the destitute. People to whom everything was due and those who dared not ask for anything, driven about as they had been from country to country for years. Yet, with a certain solidarity in distress, the common sense of having together escaped the horror. But also remarks about the Jews and the riff-raff of Europe. And the rumors that were spawned, and spread, continually inflated and renewed by the crowd's inactivity and unrest: rumors of bombings, of a German invasion, of spies planted among us, of an English refusal to admit more refugees. All the reports that bubbled, only to ebb away and die out in the face of the imperturbability of the British civil servants. What a lesson was English organization, functioning perfectly without tickets, without surveillance, with no precautions against gatecrashers and cheaters. Here people were trusted!

Days of uncertainty and anxiety. Fortunately, there was Roger D. We were always together. We constituted a group apart in the middle of the fifty Frenchmen who themselves were a group apart: mainly young men, not yet drafted who, with no family to take care of, could leave France. A heterogeneous group, both socially and politically. "There are a few too many Jews here," said one of our companions amiably. Roger, who was nearby, raised an eyebrow. "Oh," said the other, "I'm not talking about you, of course. You must be the one who's most upset." But as Roger kept on scowling at him, he did not insist.

Having little else to do, Roger and I talked endlessly. We vented our mutual rage against Hitler and Pétain. Our main concern was to justify our coming to England and to build up reasons for hope. We carefully avoided dredging up the past, Paris, our families. But each of us knew that, in an emergency, the other would be there, to talk and, above all, to listen. Together, we had made mud pies and played marbles in the Parc Monceau. Then, the same classes, with the same teachers and the same pals. Afterward, we had gone in different directions: he to business school, I into medicine. I now found him emerged from his chrysalis, matured, hardened, his body well muscled, his jaw line emphatic, but still the same light of tenderness and humor in his eyes. His strength was his determination. To set a goal and stick to it: that is one of the rarest virtues. I feel uneasy with persons when I don't know what they expect of life. Often, they themselves don't know.

After a few days, cleared of all suspicion, we were freed from the transit camp and directed to Olympia Hall. A ghastly, cavernous place. A sort of cross between the Saint-Lazare train station and the Samaritaine department store. Before the war, this caravansary served as an exposition hall for the display of products from the British Empire, Rolls-Royces or sugar cane, boomerangs from Australia or diamonds from South Africa. There, in the summer of 1940, they were regrouping foreigners, military men or civilians, who could make some contribution to the war effort. Hundreds of men of various nationalities were waiting there, walking up and down or wrapped in a blanket, even lying on the concrete floor. With nothing to do except a little cleaning, doing dishes. But mainly nonstop talk. What was the empire going to do? What was North Africa going to do? That fine army. All those Algerian zouaves and spahis: Was it possible that they, in their gaudy uniforms, were good only for show? Or for maintaining order? All these Residents-General, these High Commissioners, these Governors: Was it possible that they had no reaction to the occupation of France? One day it was said that General Charles Noguès, Resident-General in Morocco, was continuing with the war. The next day nothing further was said about it. And when news reached us of Mers-el-Kebir, where French navy cruisers were destroyed by a British fleet, with hundreds of French sailors killed, it was as if the Olympia had been struck by lightning. Many of our compatriots denounced the perfidy and cowardliness of the English. Others, including Roger and myself, considered the event unfortunate but inevitable. Obviously, the English could not take the risk of having the Germans seize the French fleet.

Discussions went on without end. Always passionate. Sometimes violent. With Roger our principal theme was: What to join? What army? What branch? On our arrival, we were determined to join a British unit. But in the transit camp, rumor had it that General de Gaulle was forming a "legion" of French volunteers. We were even visited by one of his emissaries, a young cavalry lieutenant with a sky-blue képi come to sell his wares. Arrogant, self-important, swaggerstick in hand, speaking emphatically and saying nothing, this snobbish, pretentious lieutenant made an odious impression. He reminded me of the officers my father described as his *bêtes-noires*. Soon afterward, however, this poor impression was corrected by a captain who came to the Olympia to speak and explain what de Gaulle had in mind, what the Free French forces were to be. Not a legion, but an army. Not a bunch of mercenaries on loan to England, but regular troops with regular officers. Their goal: to return French units to the battlefields; to bring French territories into the war; to have France's part in the struggle against Germany and its allies recognized by foreign countries. The next day, we decided to sign up. The few hundred French volunteers were grouped into certain rooms of the Olympia. There, one could choose one's branch. Along with Roger, I opted for the artillery. The family branch: that of my father, my grandfather, my uncles. I was merely following tradition. Thus, in the midst of all these upheavals, I could create the illusion of continuity.

So there was gunner Jacob strutting in a new uniform, parading through the streets of London. Not simply as a game. But to enter into a new skin, thicker, tougher, better suited to this new situation. Away from Paris and family, from home and medical school, I felt henceforth separated by a gap, a ditch dug by the fall of France and the death of my mother. Already, France was a faraway world, a dream world to reconquer, a sort of Jerusalem to find next year. I now knew that one stage of my life had just ended; that another was beginning under conditions incompatible with my past self, the young medical student, hard worker, and all-around good boy. My studies, my family, my love affairs,

Odile: all that must be packed away in a box to be opened only after my return to France. If I ever returned to France! But to hasten the return, to be something more than a refugee, a dead weight in a besieged England, I needed to have no other goal, no other thought, than the war to be fought. To launch myself into it with utter hardness, with utter violence. Against an enemy so brutal, so devoid of scruples, what could one do but fight with its own weapons? Let fury answer to fury; cynicism to cynicism. The young medical student needed to be strengthened, to have his muscles built up. To be toughened, body and soul. But would he be afraid under fire? Or would he be more afraid of being afraid?

I continued walking through the crowds of London, dense, calm, well ordered. Many military men. Young women in uniform, too. I went down the Strand, Pall Mall, Piccadilly. I turned to the right in a street lined with fine stores. I was looking in a shop window when suddenly I found myself surrounded by a swarm of girls. French girls, streetwalkers, thrilled to run in to a fellow Frenchman newly arrived from home. They dragged me to a nearby pub, bombarding me with questions: "Where were you? How was it? Tell us! Me, I'm from Cahors. I'm from Nice. I'm from Nantes. Was Cahors bombed? Do you know if Nantes was destroyed? And the prisoners? You're with de Gaulle?" Cackling, buzzing, all prattling at once, falling all over themselves in their nationalism. Finally, the girl from Cahors made a sign to me. I followed her to her place. Tall, about twenty-five, with a large brown chignon, she had a too gaudy elegance, with a gruff voice that belied a certain distinction. Her preoccupation, even obsession, was to make love with a Free Frenchman. That is how she showed her patriotism, made her contribution to the Allied war effort. Once her love of country was satisfied, she set about getting me something to eat. And when, worried about getting back late to the Olympia, I started to leave, she kissed me and slipped into my hand a five-pound note, apologizing for not being able to do better. Embarrassed, I refused. She insisted: "Everybody knows soldiers don't make a bundle." She didn't want to hear my protestation. I left with the five pounds. Confused, at first. Then rather proud of this adventure. I was a pimp! A good start to hardening the heart.

On the military base we were sent to, near Aldershot, on the outskirts of London, I quickly lost my rank of gunner. Owing to a shortage of doctors, medical students were assigned to the medical corps. This went against both my resolution to make war with all possible violence and personal risk and also against the desire of Roger and myself to stay together in the trials to come. I did my best to bend the rules, but in vain. It wasn't up to me, I was told, to decide where I would serve most usefully. So I found myself in a barracks with some twenty soldier nurses and medical or pharmacy students. At this base in the English countryside were stationed the three or four thousand men who, in July 1940, made up the Free French forces. The bulk of this troop consisted of units that, coming from Norway where they had fought hard, had not wanted to return to France with their commander, General Emile Béthouard. Namely, a half-brigade of the Foreign Legion and elements of a battalion of light infantrymen, to which were added men who, despite the vigilance of the Pétain government, had arrived either singly or in small groups from France or North Africa to gather together in London.

Days of waiting, of preparation. Time divided between the infirmary and the exercise ground, between injections and shooting, between the smell of powder and that of ether. Little unforeseen, few surprises. One morning when I was on duty at the infirmary, I was called to a barracks in regard to an accident. I found a man lying on a bed. A junior officer, with a bloodstain at one temple, a revolver next to him. Dead. His roommate, a short, keen-eyed sergeant with large ears, had returned from the showers to find him thus a few minutes earlier. The next day, the sergeant and I ran in to each other at the coroner's, where we had gone to testify. "When I got back to the room, Your Honor, the man was dead." I was to see the sergeant again twenty-five years later at the Collège de France; his name was Raymond Aron, later to become a famous political philosopher.

Saluting. Standing at attention. Walking in step. Making a

half-turn, going right or left. Making the morning rounds at the infirmary. Shooting a gun. Vaccinating the young recruits. Demonstrating, reassembling an automatic rifle. Faster. Doing an inventory of the English first-aid kits. Walking ten miles. Attending courses in minor surgery and first aid. Presenting arms. Brief outings in the evening with Roger to the pub in the next village. The quick return, before being swamped by the Canadians and Australians from neighboring camps given to drunken brawls. Everything that in normal times, in a normal army, makes military training unbearable—the ascendancy of the shoe brush over thought, of the warrant officer over the philosopher, of the drill over independence—had disappeared in this troop of volunteers. First because, having enlisted of their own free will, they accepted a certain form of discipline and authority. Then, since the noncommissioned officers were also volunteers, they were not trying to transform their men into circus horses or strip them of their personalities. This training was cleared of excesses and the superfluous. Only the necessary remained.

At the Aldershot base, those who spoke English read the newspapers to the others. But without giving up our outrage, we hardly talked about politics at all. As though we were ashamed of it. As though it was a game that had gone out of style. Each person's inclinations and myths could be detected, however, in the way he denounced the people who, in his eyes, bore the responsibility for the calamity: whether politicians or military men, the Popular Front or the Maginot Line. Once or twice a noncommissioned officer came out from the regimental headquarters in London to speak about the war and France, and to comment on events. After which M., one of my barracks mates and a first-year medical student in Rennes, fulminated against what he called "the sermons of the street urchin with pips." With his square head, his close-cropped hair, his widely spaced eyes, M. explained in his deliberate way: "It's fine with me if he comes to tell us what's going on. But telling me what to think, no! I'm here precisely because I don't let anyone tell me what I ought to think." To be able to think for himself, he and a pal had set out from a Breton port in a small sailboat. Two days later, they were lost. They had drifted for some time without food or drinkable water. *In extremis*, they had been picked up by a British freighter. M. didn't want anyone to talk to him about politics. For him, it was the source of all our woes: "the great crock of shit." An opinion held by many of the medics. Except for a few fixed on their high horses, entrenched in their faith. Two or three, no more. The eternal fanatics, of the Right and of the Left.

For me, this time of violence and uncertainty finally reconciled what I had long considered irreconcilable: the nationalism inherited from my grandfather and the rather naïve socialism of my father. The meeting of simple truths too long opposed: France's rights now mingled with human rights, and the love of country with that of freedom. Not only for me. One afternoon when a high fever was keeping me in bed, I saw two soldiers in combat dress come into the room, carrying on their backs outlandish sacks taller than they were. One man was short and skinny, with a hairless face; the other, tall, fat, and mustachioed. Laurel and Hardy. Abbott and Costello. They chose two empty beds, at one end of the room, while pursuing what seemed an interminable discussion. In undertones, at first, which came to me as a distant murmur without keeping me from dozing. Then sudden shouts. Jokes. Laughter. Memories of the Norway campaign. Clever tricks played on the quartermaster. A system for wangling better chow. Reminiscing about their civilian life. Their love affairs. Laurel was a bookseller in Paris; Hardy, a factory worker in the working-class suburb of Charenton. Then the voices died down again, only scraps of sentences, words, reaching my ears: "fascists . . . traitors . . . Moscow . . . contradictory resolutions . . . Stalin's trying to introduce nationalism . . . the Soviet army . . ." I strained my ears. Both members of the Communist Party, they were still wavering about committing themselves to de Gaulle. What did we know about him? Aristocratic hustler? Aiming to be Caesar? How not to look, in the eyes of their comrades, like traitors, revisionists, followers of Bukharin? And wouldn't fighting Hitler be, in a sense, fighting his

Soviet ally? Bloody hell! The German Nazis are in Paris and the Italian fascists in Nice. What good would it do to return to France under the Nazi boot, with the Communist Party shattered and dispersed, in hiding? And, sooner or later, the Soviet Union will enter the game to crush Nazi Germany. It is the workers' duty to reject the armistice, to fight Hitler. Wherever they are, they must consider themselves combatants. And the defeat of fascist forces will be the victory of the proletariat. Without a directive, without a tie to a party apparatus scattered to the winds of persecution and defeat, these militants were forced to make their decision alone, to invent alone a line of action consistent with their beliefs, to forge alone their new ideological platform. The reality of the war temporarily eclipsed the abstraction of the Party's watchwords; and the struggle against the Nazi dictatorship, that for the dictatorship of the proletariat. Even though the Party remained pacifist up to Hitler's attack on Soviet Russia in June 1941, there were, from July 1940, communists in the Free French forces as in the Resistance.

When General de Gaulle's next inspection was announced, the whole camp was swept with a gust of excitement, a tremor of curiosity. Most of us had never seen or heard the leader of the Free French. Some recalled a photo of de Gaulle that had appeared when he was vice-minister of war in the government headed by Paul Reynaud, before that of Pétain. But we knew mainly the tract posted on the walls of London: "France has lost a battle, but she has not lost the war." And, then, there was the name "de Gaulle," which rang like a challenge. A program. A bugle call. Like a charge of the tanks whose strategist he had become. During the walk across London following my enlistment, I reverted to one of my childhood habits. As of old, I began to ruminate to the rhythm of my step, on the theme of this name: *Gaulle, Gaul, goal, Gogol, Gaugaulle, Goménol, Goth, Gotha, Gothic, Golgotha, Golgothic, Gaullegotha, Gaullegothique.* So I was expecting some kind of Gothic monument. It was indeed a very Gothic personage that I saw when, an aide de camp at his side, the general strode before the assembled troops. The bugles. The flags.

The men at attention. It was France itself standing erect in this corner of England. My spine tingled.

A short speech by the general. An impressive figure. Enormously tall, with a huge nose, hooded eyes, head thrown back. Standing with his legs slightly apart, he had the majesty of a Gothic cathedral. The solidity of a Gothic pillar. With slow, awkward gestures drawing Gothic rib vaults, Gothic arches and vessels, Gothic portals. His very voice, deep and staccato, seemed to ricochet under the vaults like a choir at the back of a Gothic nave. He spoke. He fulminated. He thundered against Pétain's government. He stated the reasons for hope. He prophesied. He shook the world, the armies, the forces, the peoples. He described the coming stages of the war; the difficult moments; the final, ineluctable victory. He described the need for the French presence, for French troops on all battlefields. He promised us fights, victories. *The* victory. Then the general strode off. "I never saw anyone like him," said Roger that evening. He had the same impression I did. The impression that de Gaulle was beyond any doubt the man for the situation. The impression that to make war, to participate in the reconquest of France, we had found the right address.

1 September 1940. At sea on board the *Westernland*. The night before, the few hundred volunteers who formed the bulk of the Free French had sailed from England with General de Gaulle, headed for a secret destination which we all knew was Dakar, the capital of Senegal, then a French colony. Though the sea had been calm since we left Liverpool, I felt slightly unwell on this ship. Not really seasickness, but a base of permanent nausea. After my inspection of the infirmary, I had spent an hour in a smoke-filled lounge watching the Foreign Legion's noncommissioned officers gambling their entire pay in a round of poker. I needed air. I went out on deck.

Over the top of the railing I watched the sea slip away along the side of the ship. Now, each turn of the propeller was taking us a little farther from Europe, a little farther from France. What changes in less than three months! From medical school to this Marching Company, an infantry unit to which I had been assigned as auxiliary medical officer shortly before sailing. Thinking of my fellow medical students who had probably resumed their studies and hospital work, I had the strong sense of being where something was happening; of being where the action was; of not being out of the game, as were most of the people back in France: in short, of playing a role, however minor, however modest, in the party that would decide the fate of the world. "For a thousand years," Hitler had said.

Alone on the deck, I gazed at our sister ship, the *Penland*, which was proceeding half a mile away under the guard of an escort vessel. And also at the dark line of land on the horizon. Suddenly behind me, a low voice, a bit husky. "That land over there, what is it?" Turning around, I found myself facing the Gothic cathedral. A beret on his head, his hand shading his eyes, more immense even than at Aldershot, de Gaulle was examining the horizon. Snapping to attention, I stammered: "I don't know, sir. I would guess it's the coast of Ireland." "Yes," said the general, "that must be Ireland. It's said to be beautiful. But we'll have to wait for another occasion to visit it." He took three steps on the deck, then came back to plant himself before me. Glancing at my stripe on a red velvet background. "You're a doctor?" "A medical student, sir." "Oh!" A pause. "What unit?" "Marching Company, sir." "Aha!" The arrival of his aide de camp put an end to the interview.

22 September 1940, aboard the *Westernland*. Shortly before seven, I am summoned to see Captain D., who is in command of the Marching Company. For some days, since putting into port at Freetown (in Sierra Leone), we have been sailing with a powerful British fleet. The landing at Dakar is predicted for tomorrow at

dawn.* The twenty combat groups who make up the Marching Company have received final instructions. Each of us knows where to go, how to behave. In principle, things should go fairly smoothly. But we have been notified of the recent arrival of two Vichy cruisers that the English let sail through Gilbraltar. When I present myself to the captain, he is rather nervous. Thin, black-haired, dry-skinned, he is a reservist who commanded a company on the Maginot Line. In civilian life, he is the principal at a school for retarded children. He keeps agitating his right elbow which he cannot unbend owing to an old fracture. Seeing me, he frowns: "In our planning, we forgot to think of the planes at Dakar's airfield. We must have a group to stop the planes from taking off or landing. The only man I have is you. You will take your nurses and your stretcher bearers. You will go to the airfield with Lieutenant P.'s group. And there you will disable the planes." Stupefied, I look at him open-mouthed. Finally, I ask: "And what should I do to keep the planes on the ground?" Furious, he shakes his elbow: "I don't give a damn! Take a sword or a pistol or anything at hand and flatten the planes' tires." "Yes, sir." Then I withdraw to go announce to "my troops" the glorious mission entrusted to us.

10 October 1940. After the failure at Dakar,† the Marching Company has sailed south to the port of Pointe-Noire. A train inland to Brazzaville on the Congo River. There, in this part of French Equatorial Africa that had rallied to de Gaulle less than two months ago, an incredible reception awaits us at the station. The square is teeming with people. Blacks, whites. Civilians. Enlisted men, officers, generals. Dominating everything are the military bands with their brass and their drums. A moment of

* Senegal had remained under the control of the Vichy government. The aim of this expedition to Dakar was to have it join de Gaulle's forces.

† Despite several attempts at landing by the Free French forces and some gunfire exchanged between the British and the French fleets at Dakar, the Allied forces withdrew and Senegal remained under Vichy control until 1942.

intense emotion, even of gooseflesh, when we begin to parade in the streets with the battalion of African infantrymen to the tune of "You'll Not Have Alsace and Lorraine," a famous marching song from the First World War. All my upbringing, family patriotism, tradition mounts to my throat. The warmth of this welcome gives me a keen sense of no longer being as lost, as orphaned as in England; a sense of having found a place on earth that, despite the distance, is a bit of France; a sense of being at home here.

31 December 1940, evening. Libreville, Gabon. Feeling low. I went to this little bistro where soldiers and sailors come to drink and dance on Saturday night. A particularly grim place. With a single record, a tango, played over and over. To the point of satiety. That only increased my homesickness for Paris. Where is my father? What is he doing? What is he thinking? And my friends? Have they quietly gone back to the university? And Odile? Has she finally married the guy she was engaged to? And maman? She's better off not having seen all this. I walked for a long time by the sea. The governor's palace was all lit up. Going by it, I heard laughter and singing. At midnight, small torpedo boats anchored in the harbor sounded their sirens. I returned to the native village. It was buzzing with life and whispers in the night. What would 1941 be like? Not much chance of clearing things up by then.

31 December 1941, evening. Fort-Archambault, Chad. There is an officers' reception at the quarters of the colonel in command. Neither officer candidates nor auxiliary medical officers like myself have been invited. Furious, two of the officer candidates, D. and B., decide to play a practical joke. At midnight, they will simulate a commando raid on the colonel's residence. I raise a few objections about this plan which does not seem in the best of taste. But, not wishing to be called a wet blanket, I am forced to go along with them. At eleven o'clock, B. and D., who have had

too much punch, go looking for two submachine guns and blank cartridges. Another officer candidate and I have the job of "putting them off the track by issuing war cries in some unknown language." At eleven-thirty we are in position around the residence. Lights, music, voices, dancers' silhouettes. We begin to sneak up closer, crawling through the bushes. But B. and D. cannot wait until midnight. Each sets off a burst of gunfire in the night. Stupor. The voices fade. The music stops. Commotion. Several men come out of the residence to have a look. We just have time to slip away into the night.

31 December 1942, evening. Somewhere in the Fezzan, in southwest Libya. With all lights out, we are driving in the dark. Some days earlier, with other units from Chad, our Twelfth Company broke through the frontier to enter Italian-occupied country. After an engagement with Italian motorized troops who withdrew, our column is advancing toward the post of Um-el-Araneb which we are to surround before dawn. I definitely did not eat enough carrots when I was little, for I have trouble making out things in the dark. I screw up my eyes trying to guess at the road. One nonexistent mountain after another looms up before my eyes. I do not, however, spot the rut into which we have almost fallen. Sergeant T., who is driving the truck, makes an effort to follow exactly in the tracks of the truck that is making its way through the dark ahead of us. We talk little. One single fixed idea: to avoid at all costs the piece of rock that might disembowel the gear casing, or the soft sand in which we might get stuck.

31 December 1943, evening. Rabat, Morocco. I am on duty at the infirmary of the Second DB. We have drawn lots to determine who should do this and who should do that. I am no longer the youngest, but I'm not in luck. Too bad! On top of which, to kill two birds with one stone, I asked the dentist to extract an im-

pacted wisdom tooth this afternoon. Now, the anesthesia is beginning to wear off, releasing the pain. A pain that my boredom exacerbates no end. I am alone with an orderly and a nurse, both as upset by this sad fate as I. The others have gone into town: headed for the lights, wine, and women. Here there is nothing to do. A dull book. Is this to be the last New Year's Eve away from France? The enemy tide is now at an ebb on just about every front. But the beast is tough. And meanwhile I am bored to death. The anesthetic finally wears off completely. I am in pain. Pain over one whole side of my face. And the anesthetic has made me sleepy. Above all, don't fall asleep on duty! To stay awake, I take two pills: a stimulant, amphetamine. And immediately I fall asleep.

31 December 1944, evening. Paris. The Val-de-Grâce military hospital. A small room for two. My roommate is a lieutenant in the DB, a spahi who had his arm taken off by a shell in Alsace. I had several visitors this afternoon. My father, of course, touching as usual with kindness and little attentions. My uncle also, still very respectful of military men. And then a cousin, an elderly female busybody whom I thought I would never get rid of. None of my former friends today: all those who stayed on here, to continue their medical studies, to prepare for the *internat*,* avoid me. They do not feel at ease with me. I was hoping for a visit from Hélène, a dancer who grants me her favors. But she sent word she couldn't come. This evening, we have a better dinner, a pompously baptized New Year's Eve supper. And afterward, a treat: the lady patronesses with chocolates and presents. An ordeal! I am still in great pain. In my arm and my thigh. It has been nearly a month since they took off the second cast, the one that held my thorax and my whole right arm. The therapy for the elbow is extremely painful, and I am barely able to extend my arm. As for the thigh, I thought that it was all right. I began to

* Competitive examination for becoming an intern in a Paris hospital.

limp about with a cane. But the suppuration started up again. The surgeon wants to operate once again to get out one of the pieces of shrapnel still in there, and to scrape the bone. I will never get out of this!

10 March 1943. Southern Tunisia. A frigid day rose over Ksar Rhilane. Out of the darkness emerged a sandy, pebbly, and tortuous landscape that had not changed since the earth's birth. A world without life, where the only mark left by man was a well. A shivering desert dawn with, to the east, a pink and green sky against which in the distance were silhouetted the black Matmata Hills. Then, in the west, the pale, sandy folds of the Erg became visible, now turning to gold. That morning, we first saw the armored cars of Captain S. return. They had until then been on patrol in the mountains toward Ksar Tarcine and the Jebel Outid, covering the northern position occupied by the bulk of General Leclerc's column, known as "L Force." When we saw them arrive, distant dark points starting out of the darkness and getting bigger as the day wore on, we knew things were going to get hot. For me, this day marked the turning point of the war: the exact moment fate tipped the balance.

A day of testing. Even of trial. For up to now, Leclerc's campaigns—first, simple raids, then the campaign in the Fezzan in the past few weeks—had been directed only against the Italians. But this time we would be facing the Germans. And the German air force, which had been scouting us constantly for some days, sometimes firing at us. And German armor amassed, as the English had notified us the night before, to the north of Ksar Rhilane. For the French soldiers from Chad, it was the first encounter, or the first re-engagement, with German units since June 1940. Many of us had had to flee France; to go to England; then to Africa, the Congo; to go up to Chad; to cross a thousand miles of desert; finally to reach the Mediterranean in order to face the Nazi armored divisions at last. Once this point was reached, there was no turning back. We could allow ourselves neither defeat nor retreat. But for our group of vagrants from Chad, for the desert rats, dressed with odds and ends, driving ill-equipped, ill-armed trucks, an engagement against an armored division of Hitler's Afrika Korps represented an adventure on a new scale. A wager. A long shot.

A day of discovery. Of initiation. During the campaign of the Fezzan, I had received what is known as a baptism by fire. But the fights I had witnessed looked like naval battles: brief engagements at some remove, against forts or distant vehicles. A few cannon shots and the Italians beat a retreat or surrendered. Leclerc's performance there was, more than a fighting exploit, an excursion to the end of the world. More than the fights themselves, the crossing by three thousand men and their matériel over an impossible desert, in enemy territory, thousands of miles from our home bases. But in the lunar landscape of Ksar Rhilane, quite a different affair was brewing. For the first time, I perceived the approach of combat physically, throughout my whole body. A violent feeling. As strong and precise as hatred or sexual desire. Everything in the landscape, in the sky, in the dunes became an indistinct threat. And when, at first dawn and deep cold, the first cannon shot resounded, it seemed the moment when the war began in earnest.

A day of exaltation and fear. I no longer wondered which of these feelings would win. When the shells began kicking up showers of sand, when the noise of motors and tracked vehicles drew closer, the lighting suddenly changed. When events are so strong, so closely linked, as the sea to wind, as life to time, they are totally commanding even to the way one tries to overcome them. They carry you along. They blend instinct and intellect. It is their very violence that gives you strength. The anxiety of interrogations, the anguish of choice vanish in the struggle and the effort to survive. But very quickly I also discovered the disadvantage of finding myself without a weapon in the middle of a battle. At a time when one could finally fight, finally kill, I had in my hand a mere box of bandages. To the question, "Am I going to die today?" one can reply with a machinegun or a cannon, not with a stretcher. I envied my buddies around me, clutching their weapons, a twisted smile on their mouths.

A day of waiting. In silence and immobility at first. Everything in our position was buried: men, cannons, weapons. Hidden in foxholes. Covered with netting. Leclerc's order was definite: No movement. No noise. Not a shot before the enemy was there, close by, within reach; before we could be sure of hitting our target. We had to let the enemy look for us, explore our position whose layout it didn't yet know. About seven o'clock, we saw the first German armored cars emerge from the dunes, northeast of Ksar Rhilane. Then we heard explosions, a bit muffled, a bit scattered. Dull explosions, as if from beneath the earth. Then a furious rumbling, like a sharp intake of breath, the suction of air by a shell that passed by to explode far behind us. Then, very close, two extremely violent bursts, the sand spurting. A burst of machinegun fire in the distance. Still another explosion. The enemy was probing, groping, trying to get us in its sights, waiting for our response to locate us. In the interval between explosions, the silence hung suspended, panting. No one budged. The men stayed flat on the ground. Buried, camouflaged, L Force remained frozen. Over there, in front, I spotted the heads of our African soldiers just above the foxholes, immobile. The bald head of Lieutenant P. riveted to his field glasses. Crouching next to me, his eyes level with the ground, Toubalba, the huge nurse from a Sara tribe, muttered: "But when do we get to fire?" The German artillery continued to batter our position. In front of us. Behind us, in the Erg. To the right. To the left. At random. The position stayed silent. Then, in the distance, there arose a muffled hum. A deeper and deeper vibration. More and more strained. Like a crescendo that filled the sky with roaring, and tore the desert air. Then some forty planes with black crosses blasted across the sky: Messerschmitts and Stukas, nosediving, firing, bombing our emplacement haphazardly before sweeping back north. The German armored vehicles were now being deployed ever closer, huge mechanical insects, waddling, maneuvering by fits and starts, in the battle as in a parade. Behind them, the grenadiers advanced in short bounds, shooting in random bursts, peppering our position without yet knowing our emplacements. "What are we waiting for?" repeated Toubalba. And, as in response to his question, Ksar Rhilane was instantly ablaze. The enemy had arrived, very close, within reach. Suddenly, everything in the position that could fire did fire. Finally unleashed, cannons of all kinds, machineguns of every caliber, mortars, spat with full force. Several German vehicles were aflame. Thus informed, the enemy began to adjust its fire. Shells came closer. Bullets ricocheted, whistling on the sand. In my greatcoat, which I had thrown to one side of our foxhole, Toubalba showed me where two bullets had hit. Tanks were approaching with a roar. At that very instant, the sky and the earth were once again filled with an increasingly strident howl. This time, the Royal Air Force! Some forty Spitfires and Hurricanes. Half at high altitude to monitor the sky. Half flying low to strafe the German lines. Strong explosions shaking the ground once again. Heavy black spirals of smoke rising behind the hills to the north and east. Final passing over by the British fliers, followed by Free French eyes conscious of owing more than they could ever repay. The sky was emptied. The German armored divisions had fallen back. In the distance, munitions continued to explode. Some men in the company were wounded—several slightly, and I was able to dress their wounds, but the two more serious cases were evacuated to the field hospital.

A day of uncertainty. Of sudden reversals. Of anguish. There was hardly any chance the enemy would let it go at that. That it would admit defeat. Twice, at noon and then again at four, the Nazi tide returned to the attack. Twice, the enemy circled our emplacement, looking for weakness, infiltrating little by little. The first time, to the south; the second time, to the east of our position, always following the same scenario. The shrill eruption of the German air force, of Stukas nosediving, sirens shrieking, shooting, dropping strings of bombs whose blasts shook the earth like hail. Artillery bursting and pounding our position at every point. The approach of armored cars and tanks followed by the infantry sticking close to the explosions of their artillery, gradually penetrating our lines, despite the raging response of our batteries demolishing enemy tanks and trucks without, however, checking their advance. And then, as by a miracle, just as the situation was becoming critical, when the enemy had succeeded in cutting into our position which now included the dead, the wounded, and vehicles on fire, a roaring suddenly arose from the Erg, ending all movement at Ksar Rhilane. The British fighter bombers loomed up flying close to the ground; diving at the panzers deployed for attack; rocketing up to dive back at the tanks; hedgehopping to strafe the formations of infantry; once again, forcing the enemy to pull back. And once again, the black smoke of burning enemy vehicles rose in the silence that followed the passage of the planes. The strange silence after the pandemonium of combat. The feeling that we had had a narrow escape. That, with each attack, the Germans had succeeded in penetrating more deeply. Deeper at noon than in the morning; deeper yet in the afternoon. The feeling that the game was only deferred; that, behind the hills and the dunes, the enemy tanks lay hidden, camouflaged, waiting for a fresh opportunity; that Allied aerial intervention could not always be counted on; that the Germans might well make good on their next attack; that the epic that had set out from Chad across the Fezzan and Tripolitania had a good chance of ending in this ditch of sand, around a lost well.

A day, finally, of victory. Even of triumph. Since morning, the two air forces seemed careful to avoid each other. At each attack, the Germans came first, to back up their armored attack. Only after they had left, when the sky was empty, did the English arrive in their turn to check the attack. Around five o'clock, after the last flight of the British planes, the badly hit Nazi lines fell back to the north. In our position, we were dressing wounds. We collected the dead. We evacuated the wounded. The anxiety, the tension of combat had yielded to a sort of weary exaltation. The sun went down on the Erg. When suddenly a rumbling of planes was heard, ever louder, once again. In a flash, the sky was filled.

First, the insignias of British aircraft loomed up from the Erg. And almost immediately the black crosses of the Germans, arriving from the south. For the first time, the two formations were face to face above the Ksar Rhilane. All activity on our position ceased. Every man got ready, eyes peeled, for the show. And what a show! What a ballet of life and death! With arabesques, ascents, darts, swoops, chases, skimming the ground. Hard, brief, clean. In a few minutes, several German planes went down in flames, trailing smoke. Parachutes opened. Cries of joy among the Free French standing to acclaim the English fliers. Masters of the sky, the Spitfires returned to fly over our position, then veered to the north in pursuit of the retreating Nazi planes. The last explosions shaking the earth in the distance. The last columns of smoke rising from Jebel Outid.

Day of youth, of renewal. After the wounds were dressed, and the last of the wounded evacuated, came the time to rise from the rubble. I felt as though I were returning to earth after a nightmare. Evening fell. Once again, silence enveloped Ksar Rhilane, all the more profound for having been so violently disturbed. Even here, men had traveled thousands of miles to rip each others' guts out. Neighbors had come from France and Germany to kill each other on an uninhabited, lifeless land. A strange land, suddenly transformed for a few hours into hell and now recovering its peace, its impassivity. In the darkness that gradually blurred the surrounding shapes, the unity of the night seemed to testify to the unity of the world. I felt a new life being born. Like an escaped prisoner who, in the evening of a long march, reaches the summit of a mountain to find a land of welcome and liberty. The universe seemed to me full and mysterious, like a young animal. Beyond this sand, beyond the mountains, there was the sea. Beyond the sea, France: so green, so full of life. And, for the first time in three years, I knew physically, in every particle of my body, that the return to France was no longer simply a dream. That nothing henceforth, and no one, would prevent us from going home. Nothing save death. "So, doc, it wasn't for today," said Lieutenant P. in passing, on a tour of the company. "Him, fool," muttered Toubalba, who did not get his drift. Seated on a chest of medical supplies, I watched my comrades, black and white, privates and officers, revert to the pace of their daily routine; rediscover without hesitation the gestures of every day. What I saw especially in them was the animal side of man, a kind of carnal tenderness in the simplest movements: preparing their rations, spreading the goatskins that served as beds. As though the species, issued from the mists of time, was once again leaving its caves to come, through the millennia of strife and effort, to continue here its struggle against cold, hunger, darkness, and beasts. A stunning victory was this day's. The vagabonds from Chad had met head on the formidable Nazi war machine. In these soldiers, exhausted and happy, I discovered a new face. A new smile. The rather myopic smile of Second Lieutenant D., a zoologist in civilian life, who theorized about this battle as he theorized about any event, whether political, military, or scientific. The sly, slightly predatory smile of Lieutenant P., who today managed not to annoy me. The sweet, mysterious smile of Sergeant C., who, in this desert, dreamed and talked only of the pastures of his Normandy. The dazzling smile of Toubalba, which in the dusk revealed wide, white teeth. And even the smile of the general. A thinner, more reserved smile beneath his little mustache, as he stopped by on his inspection of our position. But the apparent impassivity of Leclerc, dry and straight in his battledress, in his tarboosh rigged up like a képi, was belied by the fine lines around his eyes, more numerous, more deeply etched than usual. As if his joy remained internal. As if his jubilation was blazing within.

Despite my fatigue, I had trouble getting to sleep. Everything had stopped now at Ksar Rhilane. Nothing stirred. The silence was disturbed only by the final whispers of the men holding forth on their exploits of the day. And also by the tart sound of a harmonica in the distance, a sort of plaintive chant that bounced around before expiring on the sand. I walked back and forth, watching the shadows, transported by the invasion of this earthly night. How could I not feel in harmony with the majestic and shadowy outlines of these nocturnal spaces? With the constellations streaming through the desert sky? With the serenity that now enveloped Ksar Rhilane? A serenity that plunged as deeply into the centuries as into the shadows. This day, begun in tumult and torment, was ending in calm and joy. But how could it be pure, this joy that seizes man, the only animal that knows it will not last forever? How not to think that this day could have been different? As could have been different this world that surrounded us? This ocean of sand. These mountains. These stars. And, above all, these men who, sprawling on their goatskins, overcome with fatigue, were, despite the fever of victory, finally sleeping. In the distance, the harmonica was still.

October 1940. General de Gaulle's first visit to Brazzaville. Flags. Music. Parades. Reviews. Speeches. Inspections. A great reception at government headquarters. The guests line up to pay their respects to the leader of Free France. Standing in front of a tent, huge in his linen uniform, the Gothic cathedral is surrounded by the principal military and civilian dignitaries. The guests are presented one by one. He has a word with each. Deep bass voice, a bit mocking, with harsh accents. The president of the Association of War Veterans arrives. Emotion accentuates his normally slight stutter. "Ggggggeneneral, my ccccomrades and I wwwwould like to hhhave the pleasure——" "The honor, monsieur," the deep voice cuts in, "the honor. Next."

Silence. The memory of total silence. Under the noonday sun, the little post in the town of Mao, in Chad, submerged in heat. In the courtyard, not a shadow on the burning sand. Behind the white walls, their blinding light, the buildings seem dead. Nothing moves. Not even the air. Not even the soldier standing guard at the entry. Not even a dog. Every living thing is taking a siesta.

I arrived in Mao one evening in February 1942. An auxiliary medical officer, I had just spent three months in Fort-Archambault as an assistant to the garrison doctor. My posting to Mao had been decided by General Adolphe Sicé, the director of the Free French Africa Health Service, as a result of what he had called "an act of insubordination." General Sicé was one of the outstanding figures in Free France: first, because of his worldwide reputation as a doctor and researcher on tropical diseases; also by his attitude at the time of the armistice when, determined to continue the struggle, he had played a determining role in rallying Equatorial Africa to de Gaulle. Of average height, with graying hair, and a thin mustache that went with a rather predatory smile, he reminded me sometimes of my grandfather in the mixture of benevolence and energy emanating from him. Many times since my arrival in Africa, he had shown me, young medical student that I was, the warm and familiar attentions of an elder. When he came to make an inspection in Fort-Archambault, where I was then posted, he suddenly announced his decision to send me to Beirut, recently taken from Vichy, so I could get on with my studies at the medical school there. In vain I respectfully protested, explaining why I could not accept such a tempting offer; pointing out that I had not left France and joined de Gaulle to go fishing; recalling that in 1940, when I signed up in London with the Free French, I had wanted to be in the artillery but that, for lack of qualified doctors, the medical students were summarily assigned to the Health Service; asking the general to post me to a combat unit as soon as possible. Nothing did any good. The general was as pigheaded as I. I understood then that, if I did not want to find myself in Beirut next week, I would have to strike a decisive blow. I stood at attention to declare solemnly that if I were posted in Beirut, it would be my unhappy obligation to desert and join a British unit. The general started. Immediately his face closed. He frowned. His mouth tightened. Just like my grandfather when he was angry. And then came the finest dressing down of my military career, on the classic theme: a soldier must obey or there is no army; especially in a situation like ours, where it was not up to the individual to decide where he would be most useful. I knew that he was right. And I quickly realized that he thought I was not wrong. The ghost of a smile indicated that he was not unmindful of the merits of my argument. With a twinkle in his eye, he suddenly told me: "Okay for a combat unit on the next campaign. Meanwhile, you will familiarize yourself with life in the desert." In a half-hour, I had passed from one extreme to the other, from the medical school in Beirut to the post in the bush in Chad where, with my two years of medical studies, I found myself the only doctor in a territory that, though certainly sparsely populated, was as large as several departments of France.

My arrival in Mao was greeted with unmixed joy by at least one person: Captain C., the doctor I was replacing. Small, plump, with laughing eyes, he spoke with a magnificent Corsican accent. After three years in Mao, he couldn't take it any more: "Not a damn thing to do in these parts, you understand. Hunting's always the same old thing. Black women are not good screws. You've got to go Fort-Lamy to find a white woman. And to go to Fort-Lamy, you need to have leave. Then you understand . . ." I understood he had but one idea: to show me my duties as quickly as possible and be on his way. To have a good leave before going to his new post in Douala (at the time, the French Cameroons): "A real city, you understand."

The doctor's role in Mao was, first, to inspect the garrison, some three hundred African soldiers stationed on the post, which was a sort of small walled fort set on top of a small butte. The morning after I arrived, I found C. at the post infirmary, whose mysteries he was to reveal. And, first, the inspection ritual. The first soldier presenting himself complained of a headache. With a jeering eye, C. asked me, "What do you do when a soldier says he has a headache?" "I give him two aspirins," I answered, slightly taken aback. "And if he's malingering?" "I still give him two aspirins." "Yes, and inside a week the whole garrison has a headache! No, believe me, there's a simple way to verify what they say." And C. took out a pocket pendulum.* "You hold your pendulum in the right hand. You point your left index finger like an antenna toward the patient. If the pendulum turns, give him the aspirin. If not: unjustified trip to the doctor; two days in the brig."

Another of the doctor's duties in Mao was to see the health of the local civilian population. Some three miles from the post, there was a village of a several hundred people. A small dispensary received the ill of the village and some neighboring villages, as well as nomads who came for treatment. C. took me to visit the dispensary. The place was a little like a den of thieves. Under the wing of a great devil of a black male nurse, there chattered and swarmed a little world out of Breughel: the lame of all kinds; the wounded with purulent, stinking sores, enormous legs like elephants' feet; the deformed faces of lepers; old blind people; children like skeletons, their bodies covered with scabs. Enough to discourage all good will. "You can't take care of everything," C. explained. "So you have to attend to the most important thing, you understand. And the most important thing is the whores. Because of the soldiers. If the whole garrison gets the clap, you can just imagine it. No more garrison. If Vichy attacks even a little, our goose is cooked. But with the whores, don't go rooting around inside. That would age you overnight. So you take your pendulum. In the military list, gonorrhea is disease number sixteen. If your pendulum turns sixteen times, then you're sure. You take the girl out of circulation and have her washed out with permanganate. And then this is useful in other circumstances. In a dance hall. A girl attracts you. You wonder whether you can go with her without risk. You take out your pendulum, discreetly, of course. You point your finger straight out. If the pendulum doesn't move, you're fine. If it turns sixteen times, then keep

---

* A pendulum is a small metal cone at the end of a string, used by some magicians to divine.

your fly well buttoned.'' The rest of the visiting hour was made quick work of.

The last duty that C. insisted on showing me before he fled was the officers' mess, traditionally reserved for the doctor. C. summoned the cook, a strapping young man with a beaming face and a brilliant smile. "In the morning he proposes the menu for the evening meal, you understand. It's always the same thing. No use getting worked up about it. First of all, there's nothing else to eat. And, second, that's the only thing he knows how to cook." Saying this, Captain-Doctor C. vanished, and I heard no more about him.

In the great press of memories, the few months spent in Mao occupy a special place; for this mixture of oppressive heat, of sand, the isolated fort, formed a new world, a world recalling old fables, the folklore of movies about the Foreign Legion; for, in a land with neither streets nor stores nor cafés—in short, without what seems to give our cities drive and sparkle—the things of life took on a new aspect, the order of values was overturned; for, in finding myself stationed close to a border, however calm, and being suddenly invested, without help or support, with responsibilities that I judged important, I seemed finally to be beginning to play a role, however modest, in this war; for, in a land with neither newspaper nor telephone, cut off from everything and everyone, I did not and could not have any news of those, relatives or friends, whom I had known not only in France but even in Africa. Loneliness here was part of the landscape.

Like all the military, I lived on the post, where I had a small two-room apartment. For my companion I had a young cat given to me on my arrival by the battalion chief and commander of the garrison and the region. All black, the little cat was gracious and playful, with a charming habit. Every morning, I found it sitting beside my bed, waiting for me to wake up, and holding in its paws a little present: the corpse of a mouse or a snake or even a scorpion, freshly sacrificed. After the cat, I had to deal with the nurses: military ones at the post infirmary, civilian ones at the village dispensary. Among the former, a striking character, Alfred, from a tribe in the Congo: the Kouyous. He had been stranded in Mao (no one knew how) two years earlier. With very black skin, fine features, and an eye gleaming with cunning, Alfred had filed teeth. I sometimes teased him: "Alfred, you're a man eater." "Oh no, lieutenant, you are mistaken. I do not eat man," he hastened to reply. Then, after a pause, he added, with a malicious smile, "My father, he eats men. But not me!" Shrewd, clever with his hands, Alfred knew his nursing. He had a good knowledge of the local diseases and often gave me good advice. We got along splendidly. He sometimes had the sad look of one who is homesick. "Here," he said, "I don't have any friends. I don't like the food. The women don't like me. A bitch of a colony!" One day Alfred arrived at the infirmary in a state of great excitement, and announced, "Lieutenant, I gained a son!" Surprised, for I knew Alfred had been away from his village for nearly two years, I asked him a few careful questions. I learned that, in keeping with his country's custom, Alfred's absence meant that his place with his wife was taken by his brother. But if a child was born, it was Alfred's.

In the evening at the mess, under the incredible multitude of stars that dot the tropical sky, we would be four for dinner: Captain S., a colonial magistrate who had rallied to Free France from its first hour; Lieutenant P., who had fought in France and had just escaped from a garrison in Morocco; the officer candidate T., who had studied for the Ecole Polytechnique and had joined up in England in 1940. Each thought only of fighting. Each accepted only very temporarily this garrison life of exercise and waiting. Despite the moments of anxiety and uncertainty, despite the temptations to discouragement, each of us steadfastly believed in our ultimate victory. None of us doubted the justness of our cause or our coming power. But, at that time, this was still an act of faith. The rejection of defeat and of Nazism had always seemed to me to be rooted in passion, in a sort of mystical belief in France and in man, rather than in reason and in the existence, in the free world, of forces capable of crushing the enemy one day, as de Gaulle had proclaimed in 1940. The second half of 1941 had begun to bear out the general's predictions. The entry of the Soviet Union and then of the United States into the war had slightly relieved the terrible feeling of aloneness we were feeling in the heart of Africa. At the beginning of 1942, however, the situation was scarcely promising. Greece and Crete were occupied by the Nazis. Even if the German army was stamping its feet outside of Moscow, it was slipping easily toward the Caucasus and the Crimea. In the Far East, nothing yet seemed able to slow down Japan's conquests. As for Tripolitania, the swing of the pendulum then favored General Rommel, who was propelling the Afrika Korps toward Egypt. Periodically, Vichy spoke of coming to bring us to reason. Periodically, there arrived from Fort-Lamy alerts involving the border between Chad and Niger.

The threat of having to fight Vichy's French troops as our comrades in the Foreign Legion and tank corps had had to do in Syria was a cause for concern at the mess. A cause of dissension, as well, between the lieutenant and the captain who were opposed on many points. Beefy, with a rough-hewn face and grinning ears, the lieutenant said little. Behind his energy, his style clipped military like his haircut, he had a lack of ease, a fussy, falsely distracted air which kept him from looking like a pirate. Small, nervous, elegantly bald, the captain loved to discourse. In his gaze, his smile, his reserved gestures, were inscribed the ineffable mark of the intellectual. For the lieutenant, there were two camps in this war: those people who were for the Germans and those who were against them; all those who were not against were for and had to be fought. This reasoning exasperated the captain. Not that he disagreed about the basic point: if things had turned out badly, he would without any doubt have played his part, reluctantly like all of us, but without recrimination. Yet the captain didn't like the way the lieutenant was talking about this problem which, he thought, was more complex. As for me and the officer candidate T.—the "child" T. as the other two used to call him because of his too-graceful body and too-smooth face—the problem did not arise. If there had to be a fight in Africa with

Vichy's forces, it would not be any of our doing. If attacked, we would be obliged to defend ourselves, whether we wanted to or not. But being younger and of junior rank, we were scarcely listened to. Other evenings, other subjects of discussion. Politics, for example. Both reactionaries, though for different reasons, the captain and the lieutenant could debate for hours at a stretch to decide whether, when the war was over, it would be desirable to re-establish the Third Republic or to set up a Fourth. "Don't count your chickens before they're hatched," the "child" T. and I retorted, all ready to concoct a Fifth Republic, even a Sixth. But such inconsequential palaver was needed to while away the sweltering evenings spent cooped up, sealed in, in this little post in Chad. For, at bottom, we were all in accord. Neither the monotony nor the routine dampened our enthusiasm. We wanted to fight. All four of us. And on at least one point there was general agreement: to do everything possible, as quickly as possible, to rejoin a combat unit.

My medical activities were divided between the fort infirmary and the village dispensary. Once over my initial elation at finding myself my own boss, I felt the anxiety of being isolated; of being without support or counsel; of having not only to make diagnoses that no one would confirm or correct, but also of making decisions about treatment. Decisions, moreover, that were highly restricted, in accordance with the limited arsenal of medications at my disposal. Only cases of dire urgency could justify a request for evacuation by air to Fort-Lamy. And not for just anyone! Happily, only once was I forced to have recourse to this expedient, for a noncommissioned officer stricken with appendicitis. My role was to make do. Alone.

Twice I had an opportunity to put my taste for surgery to work. The first time happened the day they brought to the village dispensary a man of about fifty who had fallen from a roof and broken a femur. According to the practice of the time, as I had learned it in the hospitals of Paris, the thigh should be immobilized in a system of splints and traction. It took me an entire day of nonstop work with the nurse to rig up, out of sticks and string and pulleys, a traction apparatus, worthy of a surrealist drawing. Silent, the wounded man regarded our efforts with interest, from time to time grimacing with pain. Finally satisfied with the position of the thigh, I had the nurse explain to him that, after six weeks of immobility, he would not even recall what had happened to him, and could prance about as before. I returned to the post very pleased with myself. Arriving the next day at the dispensary, I noticed that the nurse had an odd smile. During the night, the patient had cut the strings that were annoying him. So I had to start all over again from scratch. I decided to change tactics and put on a cast. I spent the afternoon preparing a magnificent cast to encase the pelvis and the broken leg. Through the nurse, I explained that the patient had to remain in the cast for five or six weeks, and then we would take it off and he could prance about as before. The next day, the nurse again had the same little smile. During the night, the patient's parents had come to visit. Because the cast was annoying him, they had cut it. Unfazed, I made another cast, this time so thick that it was impossible to cut without special tools. The next day, the nurse's smile was still there, but the patient had disappeared. Finding him badly cared for, his family had taken him home.

My second opportunity to play surgeon was more dramatic. One day two men in a coma were brought to the dispensary. They had stolen cows in a village. Soldiers had caught them, and herded them in with blows of their rifle butts. When I arrived, one had just died; the other was in poor shape: dents in the skull, very irregular breathing, probably meningeal hemorrhaging. The one chance was to open him up, find the source of the bleeding, and stop it. If not, certain death very soon. I had seen trephining done twice before. I had noticed the necessary equipment while going through the dispensary's resources on my arrival. For a few minutes, I hesitated. It was a question more of nerve than of technique. But after all, why not? Not to intervene meant letting the man die. I was lucky. Without too much difficulty, I removed a portion of the skull around the dented area. Beneath, I found a hematoma with a blood vessel punctured by a splinter of bone. I tied up the vessel. The next day my patient was breathing normally; no fever. The day after that, on arriving, I again met the nurse's strange smile. My patient had regained consciousness the evening before. That night he took off. I have rarely been so proud of myself.

On the military side of things, it was calmer. Only minor complaints. One experience, however, was interesting, though not medically. Among the troops belonging to Mao, there was a "Nomad Company," some one hundred fifty camel-borne soldiers who patrolled the Niger frontier. It was the doctor's duty to go every two months to look after them. To get from Mao to the border, I had the choice of three means of locomotion: in a truck, which seemed banal; on horseback, which went beyond my equestrian abilities; finally, by camel, which seemed to me both noble and better suited to the circumstances. Naturally, I knew no more about mounting a camel than I did a horse. But no disgrace was attached to not knowing how to mount a camel. So I borrowed a copy of the *Camel Corps Manual*. For several nights running, I studied the ways and idiosyncrasies of the camel (actually, as everyone knows, the African dromedary, with one hump). I learned how to feed it, treat it, flatter it, talk to it. Above all, to drive it by delicately holding, between thumb and index finger, the cord that went around its muzzle and served as a halter; by pinching it gently on the neck between my big toes; by murmuring the proper signals to it—*Tch Tch Tch*, to get it going, *Ts Ts Ts*, to make it go faster, *Tt Tt Tt*, to make it halt. After cramming the manual for three consecutive nights, I at last felt prepared to tackle a camel. Alfred went to find one. A short time later, he came back with a boy in a *jellaba* pulling a camel at the end of a rope. We got the camel to kneel, which was no simple matter. It was written in the *Camel Corps Manual* that, to get in the saddle, you first had to place one foot on the camel's neck, and then pull yourself into the saddle using the traditional cross of Agadès in front of it. It was not written in the manual that putting a foot on the camel's neck was a signal for the camel to get up. Hardly had I put a foot on its neck than the animal got to its feet. Thrown off balance, I just had time to get down on the ground on the other side of its neck.

Getting the camel to kneel once again took time and shouting, as much on the part of Alfred and the camel driver as of the camel. Only after a good half-hour of being roared at did the camel deign to go down on its knees. Forewarned this time, I slipped into the saddle just as I placed my foot on its neck, and clung desperately to the cross of Agadès. The camel rose again with great dignity. In accordance with the rule, I held its cord between thumb and index finger, while palpating its neck between my two big toes. I uttered the customary *Tch Tch Tch*. My accent must have been good, for the camel took off. Majestic, we rode out of the post.

All went well for the first half-mile. I even felt a certain pleasure swaying to the rhythm of my camel. Until a nasty little dog came out of nowhere to bite the camel on the foot. The camel did not take to these familiarities. Irritated, it began to trot. I felt as though I were on a boat in an angry sea. Braced on my cross of Agadès, I rang the changes on *Tt Tt Tt* to calm it. Nothing worked. Suddenly there appeared the carcass of a sheep devoured by hyenas. The camel took fright, rearing up and going off at a gallop. It was the apocalypse. I shouted *Tt Tt Tt* at the top of my lungs. In vain. I ended up by pulling so hard at the cord around its muzzle that the camel was gradually forced to turn its head around. So we found ourselves nose to nose, so to speak, staring into each other's eyes. The camel continued its mad race. When I saw a grove of thorn bushes looming ahead, I judged it prudent to let myself slide into the sand, where I rolled while the camel disappeared behind the thorn bushes. I made my journey to the nomads by car.

Even though the months I spent in this small post taught me little about medicine, they plunged me into a new and fascinating world. Distractions were few. Hunting was a bore. The library amounted to a few police thrillers, the complete travel writings of Henri de Monfreid, and some novels of de Maupassant, notably *Le Horla*. I came to spend a good deal of time at the dispensary.

Talking with the nurse and the patients. Taking an interest not only in their ailments but also their histories, their lives, their hopes, their woes. I did not speak the language, or rather languages, for this corner of the desert, north of Lake Chad, is a crossroads, a place of passage for migrations; it harbors varied populations. For any exchange with these Africans, I had to rely on the nurse, who served as interpreter. But I had only limited confidence in him, suspecting him of telling me what he thought I expected to hear. I made an effort not to see the life of the Africans too much through my eyes of the average little Frenchman, of the colonialist accustomed to referring everything to his orbit and contemptuous of everything he judges outside of it. I tried to turn around relationships, to look at things from what I took to be the African viewpoint. A difficult experiment. First, because in the situations I found myself in, my relationship to the Africans was necessarily superficial, filtered as it was by what the nurse said. And also because my chief concern was still my obsession with the war, with Occupied France, the *idée fixe* of liberation. Also, I scarcely had the energy to overcome incomprehension, the misunderstanding that separated Europeans and Africans. One of my biggest surprises, and most intense indignation, came the day when, for the first time, I examined the prostitutes, "rooting around inside," despite the advice of Captain-Doctor C. The genitals of most of the women had been mutilated. The clitoris and the labia minora of almost all had been amputated. I could not accept the principle of female circumcision, the rite, the arguments about hygiene. Several tense discussions with the nurse did no good; neither of us could convince the other. When, one evening at the mess, I asked Captain S., the liberal magistrate, how we could permit such an abomination, he was annoyed and gave me an answer worthy of Pontius Pilate: namely, that we did not want any trouble about a matter that neither concerned nor inconvenienced us.

The political and religious leader in the area was the Sultan of Kanem, an amiable old man of imposing appearance and a grizzled beard. Always escorted by several horsemen and always dressed in white, he got about on a horse fitted out in red and gold leather. He had, at the end of the village, a large house called the palace in which, as word had it, he confined his many wives. When a minor epidemic of meningitis occurred, the sultan came to ask me to vaccinate himself and his wives. He refused to bring them to the dispensary and demanded that I come to the palace. It is not every day that one has the chance to visit a harem! I immediately accepted. Exhilarated, I let my imagination run wild. At the appointed day, I arrived at the palace with the nurse and the vaccination equipment. We installed ourselves in a little room, across one side of which was drawn a red curtain. I prepared to greet a succession of beauties like a parade of models. What a disappointment when the red curtain parted to reveal a second curtain, this one black, pierced by a hole some eight inches in diameter. It was through this hole that, one after another, shoulders presented themselves for injection. All different kinds of shoulders: fat or thin; light or dark; young or wrinkled; masterful or docile. To thank me, the sultan invited me to lunch, a men's lunch, of course. But neither the different kinds of couscous nor the *méchoui* (roast lamb), nor even the brochettes of roaches crunching between my teeth could alleviate my disappointment. For a long time, I sculpted in imagination the bodies that must extend out from those shoulders. For a long time, to put myself to sleep I counted not sheep but shoulders.

It was in Mao that I heard the terrible news of the death of Roger D., my old friend, my brother. He had been part of the expedition to Dakar, with a detachment of artillery. In Africa, we had been separated at first: he remained in Brazzaville while I went to Gabon with the Marching Company. Then, when I was posted to the garrison infirmary in Brazzaville, he was still there and taking the officer-training course. For two or three months, we saw each other often. I had found him, as always, strong and sweet, cynical and tender. His least remarks amused me. Above all, he had a gift for saying, with sardonic exactitude, what I dared not say, even what I dared not think. In this military life, in his school for officer candidates, he found a kind of physical flowering, a harmony that confounded me. Despite the somber hours of the spring of 1942, he had an unshakable faith in our ultimate victory. How could I imagine, when I left Brazzaville for Douala shortly before Hitler's invasion of the Soviet Union, that I would never see him again? Shortly after my arrival in Mao, Roger was appointed a junior probationary officer and posted to Moussoro, north of Fort-Lamy. We were a scant hundred and fifty miles apart. I was trying to figure out how we could get together when I learned of the tragedy. Near Moussoro, the men of one tribe had stolen cattle from another. With a detachment of soldiers, Roger had been sent to retrieve the cattle. Spotting the fugitives, he set off in pursuit, alone on horseback in front of his men. It was then that, waiting in ambush behind a thorn bush, one of the cattle thieves heaved a spear that went straight through Roger's heart. For a long time, I could not bring myself to believe that Roger was dead, that he had died in this fashion. That he, so clear-sighted and merry, so full of life and generosity, so eager to fight for his country and the dignity of man, should end up this way in a tribal row. What a friend he had been! What a presence he has kept in my memory! With what persistence! Among my contemporaries, few are the faces that have remained so sharp.

In June, a thunderbolt came out of the too-blue sky of Mao with the news of the battle at Bir-Hacheim, where a Free French brigade headed by General Pierre Koenig held in check several armored divisions of Rommel's Afrika Korps. Immense pride at first, pride at these exploits whose glory redounded to all of Free France, to all of France. Impatience, too, and exasperation at having to stay here in this back of beyond, arms crossed, while our comrades pitted themselves, and with such success, against the enemy. The Germans pursued their offensive on all fronts. Nothing, nowhere, seemed seriously to check them. In the summer of 1942, the Allies continued to descend into the abyss. Which dampened neither our morale nor our confidence. But to keep on purring in this Kanem desert while the fate of the world hung in the balance seemed to us perfectly intolerable. That evening at mess, the fever escalated. What particularly tormented us was the special situation of Free French Africa in the war. Because of its border with Libya, Chad constituted the lone front that was entirely French. Koufra* and then the raid on the Fezzan the following year had been our first offensives, the first victories won by French troops operating from French territory, obeying a French commander. But these had been campaigns on difficult terrain, five or six hundred miles from our bases, with makeshift weapons. It had taken Leclerc's skill, nerve, and luck to make these expeditions resounding victories. Then the only troops faced were Italian. But now the Germans had arrived in force in Africa. All serious planning from now on would require close cooperation with the Allied forces, notably with the Eighth British Army of Libya. On the other hand, a British offensive in Tripolitania was inconceivable without Free French Africa playing a role in the Fezzan. The Free French were a mere handful compared with the Wehrmacht. But that was their reason for existence. For the volunteers, fighting expressed the very meaning of their lives. Those who pursued the war in Africa testified to the survival of France, like the Resistance Fighters, like people executed by the Nazis. Thus when, in the summer of 1942, I finally was posted to a forming combat unit, a company preparing to fight in the desert, I realized that General Sicé had kept his word. This posting meant the next campaign. The hour of action would soon sound. I left Mao in a state of exaltation. It was the rainy season. The roads were cut off for trucks. I made my entrance into Fort-Lamy perched on the back of a zebu.

November 1942. A week ago, the Twelfth Company left its billet near Fort-Lamy and is now heading north for the Tibesti Mountains, near the Libyan border. Today, around four in the after-

---

* The main Italian fort in South Fezzan, Koufra was captured in a raid by General Leclerc with two hundred Free French in the winter of 1941–42.

noon, our truck has a breakdown. The rest of the company cannot wait for us and drives on. As soon as the motor is fixed, we continue on our way, but a sandstorm blows up, erasing the tracks of the vehicles ahead of us. Alone in this ocean of sand, we drive on, using the solar compass. As daylight wanes, our anxiety increases. Next to me, Sergeant T., normally talkative, drives with clenched teeth. We know only too well the fate of people who get lost in the desert. A fate attested to by the mummified human and animal corpses dotting our way. Behind us, the African soldiers also remain silent. Still, nothing on the horizon. Our mouths are now dry with anxiety. Night has nearly fallen. What to do? The darkness will soon force us to halt. We have water for only two days, three if we're careful. Total silence. Suddenly a fire on the horizon. A huge outpouring of relief. A smile from T. A shout of joy from the African soldiers. Within an hour we have rejoined our company, which did not hesitate, despite the planes that might be hanging about, to light a fire to signal its position. An astonishing mixture, this Twelfth Company of the Chad Regiment: seasoned colonials, career military men or planters, veterans of the battles in France and Norway, young fugitives from France and infantrymen from the Sara tribes, black warriors from Chad. Everyone integrated, homogenized, transformed into an instrument of combat through the amicable and high-handed intelligence of Captain C., a former officer in the Camel Corps who had escaped from Niger, a veteran of Koufra, a sort of greyhound with his emaciated face, long and finely muscled body, the fiery glance behind steel-rimmed glasses. A great lord of the desert with an ascetic humor, an incisive delicacy.

Miles. Hundreds, thousands of miles. On dusty roads. On blacktop roads. In the bush. On tracks. On bare sand. Halts. Departures. Villages. Oases. Eyes burning from sun and sand. The horizon iridescent with gold, hazy with dust. The long column of trucks dashing without stopping through whirlwinds of sand. Strange trucks, grinding and clanking, bristling with machineguns and small cannon; adorned with palm leaves, netting, goatskins; with, in the rear, the African soldiers, seated, clinging to the sides, their faces protected from the dust by scarves wrapped around to leave only their eyes exposed. The road is interminable. That endless desert. Life on a truck. The *truck*! Both vehicle and house. The center of life as well as a means of conveyance. The base of departure and the goal of return. Rising in the morning with the sun. Driving until eleven, until the moment when the sun stifles all motion, when the heat suspends all life. A siesta in the truck's shadow, under the chassis, in the silence of the desert. A long gulp of cool water spurting out from the *guerba*, the goatskin tied to the truck's sides. At three o'clock, back on the road until nightfall or later. The hole dug near the truck to provide shelter from the wind. The constant whispers of the African soldiers talking endlessly in the night. Fatigue and loneliness in the evening in the *faro*, the sheepskin sleeping bag, with memories and hopes, with worry and determination. The intense cold of the night when one goes on guard duty, peering into the darkness, eyeing the silence. The surprising world of the desert, as primitive, as eternal as the sky and the sea. The ocean of sand marked out by the corpses of men and animals, felled by thirst and mummified. The extraordinary variety of landscape, different each day from morning to evening. Boulders of lunar rock. Flat land opening on the infinite. Long soft dunes with feminine curves. A scree of giant pebbles. Great black *garas*, the desert cliffs with the silhouettes of fantastic animals. Long rivers of sand branching like networks of veins. Crystallizations in the form of needles. The white and gold skeleton of a barren land, populated only by mirages. Blazing horizons like forbidden kingdoms. Brief explosions of green life. Great faults of soft sand into which the truck suddenly sinks. The exasperation of the driver, who is stuck yet again. The resignation of the soldiers, yet again obliged to place boards under the wheels and push, push.

This cavalcade drove from the Fezzan to the posts in Gatroun, Murzuq, Sebha. The Italian garrisons had little spirit to resist. Barely had one palm grove fallen than the columns charged on to the next. The incredible audacity of Leclerc, who was everywhere at once, pointing his cane at the hottest point, making scenes of Homeric wrath, each time imposing his will on an enemy who, though outnumbering us, was always hacked to pieces. A fantastic epic in which puttering improvisation functioned like a well-engineered mechanism. The touching complicity of this handful of volunteers come from all over to prove that he who has crushed by tanks will be crushed by tanks. The incredible bunch of raggedy tramps in disparate uniforms, in helmet or in képi, in shorts or in long pants, in greatcoat or in burnoose, but unified less by squalor and shaggy beards than by the passion and the will to fight it out. Favored by the return of the pendulum that had, since the Allied victory at El Alamein, brought British Field-Marshal Montgomery in pursuit of General Rommel, this cavalcade from Chad to the sea, through landscapes out of the Apocalypse or the Creation, yielded the only balm that could do anything to assuage our anguish as exiles: the pride of victory. It was like a birth. Behind the opened doors of these forts, abandoned by an enemy taken prisoner or in flight, stood a new world: a world as full, as luminous as an infant's body. Life, for the first time in months, became as strong as darkness. As strong as death. And when, after meeting up with the British Eighth Army in Tripoli, we saw the sea, it seemed that in the distance, in the mists, across the Mediterranean, lay, like a mirage, the coast of France.

I saw the sense of honor, the pride blazing on the faces of the Sikh soldiers from a division of the British Eighth Army. It was in southern Tunisia, in February 1943. We were utterly exhausted when we arrived. It was night. We had ridden nearly forty hours without stopping. We had been told that the Eighth Army was going to attack the Mareth line; that L Force, Leclerc's units from Chad, would be with the left arm of the Eighth Army, along the desert; that a Hindu division would be on our right. I was so tired on arriving that I did not even open my sleeping bag, but collapsed right on the sand.

I was rudely awakened by an infernal racket. Behind us, English batteries were shooting nonstop: several hundred cannons barking and recoiling. In the night, the orange flames of the shells taking off stretched in a dotted line along the front. All around, the earth shook, as if offended by each boom. All this pandemonium was not enough to waken Sergeant T., the driver of my truck, who, lying next to me, went on sleeping as though nothing was going on. Day broke. The batteries continued shooting until about nine o'clock. Then they abruptly stopped. In the sudden, almost indecent silence, the front seemed to calm down.

Several times I had seen the Sikhs protecting our left flank. Great strapping young men with dark skin and long black beards. Impassive in their turbans, they strode the world with an imperious gaze. After nightfall, I came across a group of five men going out on patrol. Naked, heads turbaned, their bodies oiled, daggers stuck in their belts, they went off, each swinging in his hand a thin cord. These cords, we were told, were for strangling enemy sentinels. In silence.

23 March 1943. Southern Tunisia. In the early morning, the trucks unload the Twelfth Company at the foot of the Jebel Matleb. From there the enemy controls the pass to the west of the Matmatas, preventing the advance of the New Zealand armor attempting to get around the Mareth line. On their arrival, our sections deploy themselves on the foothills of the mountain. Emerging, we see, on a peak, alone with his aide de camp, Leclerc stamping with impatience. He is being fired at from nearby ridges. He responds by hoisting his cane like a rifle and shouting: "Bang! Bang! Bang!" Hardly has the company disembarked than he sends it against the obstacle. Surprised by the violent assault, the enemy is forced to pull back. The summit dominating the

valley is quickly occupied, but Captain d'A., the company commander, has been killed in the skirmish. An uncertain night, each side's positions remaining murky. The next day, the enemy tries to retake the mountain. First the Italians. Then the Germans. Several of our soldiers are wounded. Going in search of one of them, I am suddenly set upon by rifle fire and must flatten myself behind a rock. From there I see, on the peak opposite, one of our sections installed at the summit; while halfway down, a German company is trying to come up and recapture the position. Each time I try to get up, a hail of bullets spurts around my rock, pinning me to the spot. Over there, some two hundred yards away, the soldiers from Chad above, the Germans below, shoot at each other almost pointblank. The German artillery is pounding the top of the peak at the risk of hitting its own men. Suddenly, lobbing grenades, jumping among the rocks, a dozen of our soldiers charge, shouting, bayonets fixed, among the bullets whistling by from all directions. Straightaway, two of the men fall, one even rolling several yards down the slope. But the Germans give way. They take flight. A hundred at least race down the slope. Three of our men go after the fleeing Germans. One of them falls. The two others keep on running. Again, one or two blasts of machinegun fire. Then silence envelops the mountain.

As evening falls, we count the dead and the wounded. Numerous, alas! The last one killed, from a head wound, as he was chasing the fleeing Germans: Vincent D., my friend, the zoologist at the Museum of Natural History in Paris. A wonderful friend, glad to be alive, curious about everything, who in the desert nights gave us endless courses in zoology and politics. When I saw his body, his face, despite the blood, was already transformed by the calm death brings. Lines and wrinkles had disappeared along with thought. And in this mask drained of life, I saw my friend's true face spring up.

The next day, at dawn, an impressive rumbling coming up the valley wakes the soldiers of L Force, the conquerors of the day before. From the height of the mountain, we then witness, as from a box at the theater, a breathtaking spectacle. Monstrous machines, worthy of the Apocalypse, pushing gigantic blades that level everything in their path, and brandishing enormous chains that hammer the ground in front of them, advance at the same pace, peaceful and haughty, and traverse all obstacles by leveling before them dunes, rocks, wadis, minefields, and anti-tank devices. Behind, on the series of parallel tracks thus opened, rush, in a cloud of dust, thousands of other monsters, tanks of every caliber, crawling, stopping, firing, starting again. Above, waves of planes fly over in unbroken succession, diving at the enemy tanks, the antitank devices, the batteries camouflaged in the mountains. And everywhere, plumes of black smoke rising after they pass. From the height of their balcony, the vagabonds from Chad watch, wide-eyed, the charge of the British armor. How could we, in our most unbridled dreams of revenge, have imagined such revelry? How conceive of such means, such power, such assurance? Shouting with a savage joy, we take part that evening in the British breakthrough to El Hamma, toward Gabès on the Tunisian coast. No doubt now. The loneliness and the nightmare have come to an end. The door to return has finally opened.

The end of May 1943. A leave in Algiers after the Tunisian campaign. My first leave in three years in Africa. A hundred men from Leclerc's column formed a little convoy. Shortly after we entered Algeria, our truck had skidded and knocked down an elderly woman on the edge of the road. A fracture of the woman's thigh bone, and we took her to the hospital. Afterward Sergeant V., the driver of the truck, and I went together to the police station in the next town for the formalities. There a scribe sat enthroned under a portrait of Marshal Pétain. Questions. Answers. Circumstances of the accident. The driver's identity. "Have you ever been convicted?" "No, never," said V. "Are you quite sure?" insisted the scribe, without looking up. "Yes, sure." "Are you completely sure?" Nonplussed, V. stared at him. Suddenly he understood: "Oh, yes! I'd forgotten." He pointed to Pétain's portrait: "I was condemned to death by that old bastard." Impassive, the scribe continued to record.

At Constantine, five of us went off to lunch at the officers' mess. At the door of the mess, the guard took note of our open-necked shirts, our rolled-up sleeves, our shorts, and our bare knees. "You must have a tie to enter!" Furious, Captain-Doctor C. wanted to enter by force, to make a scene. We managed to calm him down. Finally, we ate in a nearby restaurant. A bountiful meal, washed down with a bountiful amount of wine. Not far from us were seated three officers from a Constantine regiment faithful to the Vichy government, one of them a major. From our table, C. watched them out of the corner of his eye. Suddenly he lifted his wineglass while staring at them fixedly. "Down with the assholes!" he declared amiably, but firmly. Bewildered, the Vichy officers stopped eating. "Down with the assholes!" we repeated in chorus with cajoling smiles. Pale, jaws clenched, the others got up. For an instant they hesitated, wanting to grab us by the throats. Then, without a word, they left.

During my leave in Algiers I stayed with Roger D.'s parents, who had taken refuge there. They had been hard hit a few months earlier. For to Roger's death was added that of his elder brother Jean, also killed in horrible circumstances. An active member of the Algiers Resistance, he had, the night before the American landing in November 1942, led a group whose mission was to occupy the main post office. An occupation carried out smoothly. Later, a squad of the Fifth Chasseur Regiment, the Algiers regiment faithful to Vichy, took up position in front of the post office. Jean D. came out to try to discuss matters with the commander of the squad. In vain. The parley over, Jean was turning to go back in the post office when abruptly, without warning, an adjutant of the Fifth Chasseurs shot him twice in the back. He died a few hours later. A spear in the heart for one brother! Two bullets in the back for the other! What a fate for these two brothers, both superb athletes; both full of life, health, humor; both Free French from the very beginning. In these trials, Monsieur and Madame D., and their daughter as well, displayed a dignity, a generosity that were the admiration of all. I had hoped they would have some news about my father, whom I had not heard of since leaving France so long before. But they knew nothing. Poor Papa! I wondered what he now looked like; what he was thinking; whether he knew where I was; whether he was in the Resistance; and, above all, whether he had managed not to be bothered too much and to pass through the net that the Gestapo was spreading over France, now wholly occupied since the American invasion of Algiers. In what state would I find my family? How many would have disappeared?

A strange atmosphere held sway in Algiers in late May 1943. A mixture of enthusiasm and intrigues, of corrupt ambition and not giving a damn. Gangs furiously confronting each other, ready for any scheme. The arrogance of people who had up to then been bullied by Vichy. The fear of those who were afraid of being bullied now. All in a setting of sun and whiteness. With the layers of history deposited by people who had never ceased detesting and loving each other, distrusting and submitting to each other, without really achieving integration. With streets where streamed Moslems and Europeans who distrusted each other; but who all, even the Europeans, distrusted Europe even more. As though Europe, formerly too small, had boiled up and overflowed, lapping at the coasts of Africa from which it was starting to ebb, casting away the men and matériel it had thrown there.

Some unexpected encounters. Friends of my family, pals from the lycée or from medical school, transformed by time and events. Like little B., once so shy and now flanked by a tall, boisterous redheaded wife on whom he kept a tight rein. F., disguised as an air force captain, but with only ten hours of flying time. T., as guileless, as nice as ever, who had played a decisive role in the battle at Bir-Hacheim. D., who still nursed the hope of soon becoming a minister of state. P., who wanted to have it forgotten that he might have become a Vichy minister of state. Madame E., in her nurse's uniform with three stripes. And then the Americans who overran everything, settled in everywhere, ready to transform the Casbah into a district of Chicago. Thanks to them, and to their protégé, General Henri Giraud, the Vichyites were still in control of the turf. Giraud did not want to welcome in North Africa the Free French troops who, after the fall of Tunis, found themselves in exile near Tripoli. But the wind changed. We were cheered in the streets. We were surrounded. We were invited everywhere. Men were constantly coming up to ask how to sign up with the Free French. The day de Gaulle landed in Algiers, Leclerc's men on leave were ordered to stand along the parade route leading to the monument to the war dead: a "hedge" of honor, but a hedge also against the risk of an attempted assassination. When the Gothic cathedral alighted from his car and began to mount the stairway, a huge ovation broke from the dense crowd that had come to acclaim the leader of the Free French. But for several days we were asked to remain available and on the lookout. Strong rumors of a coup d'état persisted.

Algiers was a decisive step on the road back to France. For us, upon leaving the desert, it was a renaissance, the return to a forgotten world. A big city. With sidewalks, stores, goods on display in shop windows, cafés, restaurants, steaks and french fries. With women, many pretty girls, their light dresses clinging to their bodies. It was also an interlude in military life. A return to a more normal life. Where one arose without the sound of bugles, without cannon fire. Where, if one was sleepy, one went to bed. Where one took baths. Where one idled in the streets: in great avenues blazing in the sunlight; in alleys so narrow as to be always in the shade, where, between linen hung out on lines to dry, one could just barely see a thin ribbon of blue sky. And everywhere, swarms of half-naked children, the color of the earth, who bickered in Arabic, who danced round and round me.

But all that was not Paris!

May 1943. The slopes of Jebel Garci, a few miles south of Tunisia. I am walking at night, alone. An uneven road, pebbly. A silence broken, here and there, only by the departure of a shell or a shot of an automatic rifle. A sky without a moon. A world of ghosts, of silent powers behind the shadows. The sound of a stone rolling down the mountain startles me. I am suddenly conscious of the war around me. Conscious of the enemy who may be crouching here in the darkness. As though a contained vigilance, an acuity of all my senses mobilizing all my nerve fibers, was insinuating itself into my hands, my eyes, flowing through my arteries. As though, suddenly, my body, on the alert, was ready for anything.

Abruptly a shell explodes behind me in the dark. I didn't hear it coming. The explosion surprises me so much that an intense, senseless fright takes possession of me. I see myself from outside. Alone. Lost in the dark. Perhaps there, just ahead, two eyes have been watching me for some time. Perhaps someone has already removed the pin from a grenade with my name on it, destined for me. I throw myself flat on my belly. Immobile, I probe the silence. Terrifying thoughts swirl pell-mell through my head. The least crackling sound tenses my body like a bow. I picture the little black hole, the mouth of a submachine gun a few yards away and aimed at me, with a finger on the trigger ready to squeeze. An access of sudden fear. Fear of the unknown, of danger. Fear of death, immediate, brutal. Perhaps an atavistic fear. Perhaps man's ancient fear of the night, of beasts, of the cunning power of evil. Sweat oozes from every pore. My hands are shaking. My breathing accelerates. My shirt soaked with sweat, my mouth dry, I remain pinned to the ground. Eyes and ears in wait, I hug the ground, I press my belly against it, unable to move hand or foot. It is useless to tell myself that this is stupid; that I am not of the slightest interest to the enemy; that no one is menacing me. In vain. My muscles refuse to budge. I remain lying there, terror in my heart. A child's panic when in the night the walls of the house creak.

And yet this was not my first night expedition. This time, however, I was alone, wandering between the lines, not knowing where either side lay, in search of one of our isolated outposts.

One company from L Force occupied, in fact, an advanced position, wedged into the enemy's defense system, separated from the bulk of our troops by an uncertain zone, little frequented during the day, but traversed at night by patrols from both sides. This company had requested a medic for their wounded. A mission for which I had been named. So as night was falling I presented myself with two stretcher bearers at the outpost. Lieutenant S., a strapping Alsatian with the face of a pirate, showed me the way: "Go about three miles, heading for that peak over there. It's a little to the right. You can't miss it." I left at night. Very quickly, I lost my stretcher bearers. I was thus walking alone toward the peak. Listening alone to the sound of my steps.

A dog barking, far, very far, away, makes me jump. There is a world still where dogs bark. A great warmth invades me. That life over there, that animal that lives and shouts wrenches me from my solitude, from my fear of death; unites me with other lives, with animals, with men who come and go around me. I no longer feel myself a little piece of isolated existence, trembling in the dark. A flare casts an orange light on the landscape. I take large gulps of air, as though to wake myself up. As though to wash away a nightmare. I look at my watch. This eternity has lasted barely five minutes. Five minutes of panic, without rhyme or reason. I get up. I adjust my medicine kit. I check my reference points. The peak ahead. The mountain to the right. The polar star between the two. I must have gone halfway now. In this moonless night, one sees both too much and too little. The fear continues, but is reasonable. A sort of prudence, of vigilance. Here and there shadows stand out. I move forward carefully. Each yard brings me closer to my goal and gives me more assurance.

All of a sudden, my blood freezes. Barely five yards away, in front of me, a human figure leaning on a bush. A German wearing camouflage and a long-visored fatigue cap. With the barrel of a submachine gun at his hip, he tracks my every movement. I have no weapon. Nothing but my medicine kit with its red cross, invisible in the night. What to do? Flee? Halt? I decide to proceed. With all the dignity I can muster, I continue on my way, trying not to run. To walk at my normal pace. I keep my eyes fixed ahead, not looking at him. But I feel the barrel of the submachine gun following me. Marching this way, I feel utterly at his mercy. But I am so absorbed in walking slowly, with dignity, that I do not even think of being afraid. I approach him. I parade in front of him. At each instant, I expect to be blasted. I go away now, as calmly as I can. The bastard does not want to kill me face to face. He is surely going to shoot me in the back. I continue walking. An oblique look behind me. Then nothing. The darkness closes in. Once again, I am drenched in sweat. My shirt sticks to my skin, and the breeze blowing in the night makes me shiver. Must hurry now. Rejoin as quickly as I can my comrades who are up ahead. But why did Fritz not fire? Why did he let me parade before him without doing anything? For more than forty years now, I have been asking myself this question. I shall never have an answer. The most probable one is that, for him, I was a quarry of no interest. That, for him, I was not worth the risk of letting his position be known. But then what does it matter!

The errand was completed without mishap. I reached the company at the expected time. A stay without comfort. The only precise memory: hunger. A hunger that devoured me and fed my dreams. Between two mortar bombardments, while the countless fleas infesting me between skin and shirt were drowsing, satiated, I dreamed of steak and french fries. I was busy with the wounded. But two days after my arrival, a mortar blast immobilized my left arm. I tried to take out the piece of shrapnel myself, but succeeded only in hurting myself. I, too, had to be evacuated at night with the other wounded. A troop of the disabled making their getaway, groaning in the silence and the night. Then we went by ambulance to the hospital in Kairouan, where the column's surgical group was set up. Over the plain rose the miraculous renewal of dawn. Forms began to emerge from the shadows. Trees. Bushes. Brambles covering stones. An abandoned tumble-down shack, with its door hanging open. An Arab jogging along on his donkey. A morning as simple, as pure as if there were no war.

On arriving in Kairouan, I learned that the Allies had just entered Tunis.

April 1944. The advance detachment of the Second DB sails from Casablanca, headed for England. A violent storm in the Bay of Biscay. I am so seasick that I would rather die on the spot than leave my bunk. Several poorly secured tanks have begun to slip. Trying to immobilize them, a soldier has gotten a finger mashed. No way to evacuate him. But the finger threatens to become gangrenous and must be amputated. Barely able to stand up, retching, my eyes floating in their sockets, I must disjoint the wounded finger while nearby a sailor holds a basin in which I can vomit between each cut of the surgical knife. A veritable nightmare. A miracle, too, for the wound heals over without suppurating.

With the formation of the Second DB, many things had changed. In Africa, the war was like the cops-and-robbers games of my childhood. There were teams. Leaders who chose their teammates. Gangs of volunteers: Leclerc's gang, Koenig's gang, each equipped higgledy-piggledy, operating with whatever means were at hand. These gangs had done wonders, through sheer force of will. They had upheld France in the fight. They had made France exist on every field of battle. With the end of the Tunisian campaign, however, this period was over. Finished was the heroic chapter of Free France, the army in rags, the vagrants skirmishing across Africa. Finished, the gangs who knocked together impossible victories. To get Free France under way, it was necessary to bypass a regular army. To get on with the war, it had been necessary to return to it. Playing a role required a change of scale, of structure, of matériel. A transformation that manifested itself when Leclerc abandoned his legendary chéchia for a true general's képi. Or when we had to throw away our rag-tag clothing for fresh outfits, all the same, all American soldiers' uniforms. From then on, the great business was the formation of the Second DB. Everyone turned to it. Leclerc the first. A stubborn, meddlesome, meticulous, irascible Leclerc. In short, an unbearable and smiling Leclerc who brought off the impossible: choosing the units that suited him; combining, despite political differences, troops from Free France, escapees from France, and regiments from North Africa; finagling an impressive amount of modern equipment from the Americans; instilling a unique spirit into this whole mixture; thus making his division into a peerless instrument of war.

Begun in Morocco, near Rabat, the training of the DB was to be continued in Great Britain. I had been posted to the Second Company of the medical battalion where I teamed up with Captain M., a highly competent physician to colonial troops. This little man was sweet and generous, shy and ardent: a sort of saint who lived for his patients and his profession. Our role was to form an emergency surgical post for one of the division's tactical groups.

The DB was billeted in the area of Hull on the southern coast of England. The medical battalion settled in the nearby town of Cottingham. A charming country village, with streets lined with identical little houses surrounded by identical little gardens. The division infirmary was lodged in a sort of large manor house. Days of intense preparation, divided between work at the infirmary and training in the field. But also evenings of relaxation. I lived in a small house with four other medical students, including Jean D. and Yves C., who had recently come from France via Spain. We had many visitors. The English ladies had greatly changed, having learned a great deal in the course of the war. They did not disdain the French soldiers, but kept their own style. One evening, Jean D. burst into my room, bug-eyed, "I've just made love with Patricia, the girl I met at the pub last night. Afterward, I asked her to take off her clothes. She refused, saying we didn't know each other well enough! What should I do?" I had no answer to that.

England was the last stop before the return to France. An indispensable stop. It was, indeed, the very opposite of Africa, the converse of Algiers or Casablanca. A soft, gray sky, rather than blazing sun and blue sky. A nation where each of its citizens was at war, rather than a country where most people were indifferent. A world where one did not see men running and cars taking off at every opportunity. Where there were no idlers hanging about in the streets or strolling through stores. Where one entered, one made a purchase, and one left. Where one did not gesticulate, did not shout, did not talk with one's hands. Where everything was shaped according to its use and to the measure of the man. Where nothing yielded to impulse, to speed, to fashion. Where effects were related to causes in a manner contrary to causes without effects and effects with a cause. Unconquerable England that did not submit to the war, but submitted the war to its habits and traditions. Adapted it to its proprieties. What a debt to her we had contracted when, alone, she stood fast against the monster!

A striking encounter near Hull: one of my old comrades from school, Jacques S. I had kept a memory of him as a boy who was well behaved, shy, reserved, taciturn, and first in gymnastics. I found a man very upright, determined, sure of himself, full of the authority conferred on him by his aviator's uniform with four stripes, covered with the highest decorations; and above all, with an aura of tragic destiny as revealed in his face deformed by scars. Having signed up with the Free French in June 1940, S. had become a fighter pilot and had commanded the Ile-de-France group of fighter planes that participated in many air battles over England and France. Shot down in flames over northern France, he had been seriously burned in the eyes and hands, but managed to parachute to safety. Nearly blind, he was able to find refuge on a farm. From there, a Resistance network had gotten him into Spain and then to England. After several operations, he had regained the use of his eyes and hands. Enough to resume training. When I saw him in May 1944, he was hoping to regain his place in the fight. He did. He was shot down and disappeared into the sea on 26 August 1944, at the very hour de Gaulle was going down the Champs-Elysées.

On the morning of 6 June 1944, when we learned of the Allied invasion, a vast thrill ran through the DB. This time the signal had been given. The gigantic machine was stirring. There would be still many highs and lows, hardships, disappointments, losses. But the die was cast. The end of the war was now just a matter of time and of technique. The important thing now would be to rebuild. To rebuild the houses, the roads, the state, the society, the institutions, the rules of the game. Everything. Everything that, in a few weeks of the spring of 1940, had been dashed to bits before the German soldiers singing on our roads. Everything had to be reconstructed. But how would this reconstruction be done? Would there be hopes and wills of sufficient stature to make up for the horror and treason? Would the return to peace, to life, have enough force to balance the inordinate amount of sacrifices and ignominies? Was the energy spent by so many in the struggle against enslavement going to carry over to less heady, less glorious, but no less necessary tasks? History had ended by resembling a dream. But once the goal was attained, would the deliverance and the reconstruction still resemble a dream?

V

NAILED TO A BED like a beetle on its back. Imprisoned as in a straitjacket. When I began to regain some consciousness of my surroundings, I found myself immobilized, my entire thorax and right arm squeezed into a cast, my pelvis and right leg into another. And too weak to move the parts of me that remained free. I did not attempt even to budge, barely emerging into the light, only to dive back into the void. A blurred time, with clearer spots in the darkness. A shapeless time, without rhythm or desire. With a base of intense pain when I touched ground again. With operating tables. With anesthetics. The rubbing against scratchy sheets. The corridors. The smell of ether. The hoists with which I was raised with difficulty. The flare-ups of fever that made the walls of my room recede. The needles in my one free arm. The heads of men in white jackets. Standing. Sitting. Whispers, secret deliberations about life and death. All I knew was that I did not know where I was. That I had lost the world.

Then, slowly, time reappeared. Days were born from endless night. Once again, amid the shadows, light shone under the door. Once again, dawn was expected, death's advances averted. I rediscovered the sense of existing. The world readjusted itself. I learned that, after the American field hospital, I had landed in the naval hospital in Cherbourg; that the doctors, the nurses, all were navy men. Then the days slipped by. And the waiting began.

Intolerable waiting that one could not get used to: waiting for the doctor and his visit; for the bedpan; for evening; for morning; for the nurse; for medication; for the X-ray machine; for the orderly; for dinner; for breakfast; for the injection; for bandages. After coming up from the grave, I returned to the newborn state: an infant clothed in a rough nightshirt, which was both too wide and too short. Who had to be wiped. Who had to be bathed. Who had to be fed. But who never saw the smile of a woman leaning over him. For the nurse's hand that cleaned and fed him with competence and authority, that tied a napkin around his neck, that brought the plate near his face, that put spoonfuls of soup and jam in his mouth, was the hand not of a woman but of a young sailor. Perhaps the only time I have regretted not being attracted to men!

The stifling heat of those summer days when nothing relieved the monotony of the hospital. Nothing, except news from the front in Normandy. One afternoon, one of the doctors brought in a radio. The DB had just entered Paris. And the radio crackled with excitement. Behind the voices of the commentators the fête resounded. By closing one's eyes, one could join the millions of Parisians whose cries of joy mingled with the pounding of tanks on the pavement. A crowd like that at a carnival: noisy, dancing, happy. Occasionally silenced for a moment by a detonation. And then quickly returning to voice its jubilation, encircling the tanks, beating sidewalks and walls with waves of happiness, covering with its clamors the hubbub of the war that was still going on. I began to cry. To my tears of joy were added those of disappointment. What rotten luck it was to find myself, on such a day, flat on my back, in this little room, alone!

But the worst of my humiliations was the return to Paris. I had dreamed of this return in every possible condition, every possible form: on foot, on horseback, by car, by ship, by plane, by helicopter, by parachute. Everything except on a stretcher in an ambulance. In Cherbourg they wrapped up me and my casts like a package. A package so heavy, so cumbersome that three men were barely enough to hoist it onto the railway coach. In September 1944, it took three days for a hospital train to go the two hundred miles from Cherbourg to Paris. Three days of jolting, of air-raid alerts, of halts at little stations or in open country. Three days of itching in my shell like an oven, amid the cries and groans of the other wounded, the only distraction and nourishment being a few slices of bread spread with peanut butter. The package was unloaded in a suburban station, transferred to an ambulance, and driven into Paris. A Paris that I could not see from behind the opaque windows of the ambulance, but that I tried to hear, to identify by its sounds. In vain. The ambulance dashed at high speed, its siren screeching. It bounced going through the entryway of the Val-de-Grâce Hospital. Great gray buildings, dirty, half barracks and half monument. A long courtyard where patients limped about in pajamas or unbuttoned uniforms. Corridors where silent ghosts slipped by. At last, the room.

A room for two. With a roommate who changed several times during my stay. The first, when I arrived, was lying between two boards, in agony. An abdominal wound that had dragged on and on for weeks. Infection. Peritonitis. He no longer looked human. He did not regain consciousness. From him, I heard only moans, a sort of syncopated rattle. He died one night, shortly after my arrival.

For weeks and weeks, I was shut up in this room, caught in the corset of my casts, pinned on my back like an insect. A small room with a very high ceiling. A box with greenish gray walls and a dirty white ceiling. To my right, the window, through which I could see only a bit of sky. To my left, the door giving onto the corridor. To the right, the world outside, the sounds from the garden, and the muffled hum of the city. To the left, the world of the hospital, with its own rhythms, its own noises: the clatter of china before meals; the squeaks of tables on wheels, foretelling a change of dressings; the footsteps announcing the doctors' rounds; the horde of doctors and nurses under the wing of Commandant-Doctor G., an abrupt little man, dry of skin and black of hair.

For weeks and weeks, I remained in this room, floating in a pool of despair, tenderness, and anxiety. Fastened on my back, I had the ceiling as a backdrop for my reveries. What attracted my gaze on awakening, what plastered itself to the back of my eyes until I slept, was the flat, precise, clearly outlined rectangle of the ceiling, with its old gingerbread, its antiquated molding. The hard geometry of the design, the brutal precision of the relief gave this surface an existence, an intensity that went beyond what sight revealed or what I would have perceived if I could have explored it with my hands. By repeatedly scrutinizing this ceiling, by examining every protuberance, opening and closing my eyes on it, I felt, with a sharpness that turned to malaise, its power to fascinate. It seemed to quicken with its own life, to become dense with inhabitants. As though the objects in the room and the murmur of life outside were drawing, on the ceiling, the map of a world, simplified but endowed with a presence all the more real for being somewhat hallucinatory, an effect I felt to the point of vertigo.

Every external stimulus was then charged with a bizarre meaning and diverted my dreaming into some channel of memory. A sudden rush of birdsong coming up to my room from the garden, the chirping of a flight of sparrows took me back to the muggy warmth of the equatorial forest during a journey through central Africa when thousands of birds whistled and chattered all night long; immediately, on the ceiling, naked blacks began to oscillate, in the slow rhythm of a lascivious dance, chanting their endless songs of love and war. As evening fell and the lights came on, there appeared on the rectangle of the ceiling great motionless shadows laden with violence and passion. One of the bedside lamps, shaped like an upside-down conch shell, traced the curve of a giant breast, the hard sweet breast of Jenny, the disturbing little pink and blonde prostitute with whom, in a town in Tunisia, I had, for two hours, forgotten the war. Or again, when my rage at finding myself in this impotent condition, tied down on my back, set aflame a desire for power and revenge, I returned to that night in Hull when I had had to command a patrol in town, charged with seeing to the good behavior of the soldiers on leave from the DB. Helmet on my head, strap under my chin, pistol at my belt, and flanked by four other soldiers, I circulated solemnly, strolling among the bars and brothels of the port. Initially furious at having to play the cop, I came to be amused by the expression of fear accompanying the salutes of the privates we encountered. An impression of brute force. Cold comfort to my feeling of impotence within my casts.

Unable to go out, unable to walk in this city of which I had dreamed so much for four years, I had to imagine doing so. Rebuilding it in thought. Rediscovering it on the rectangle of the ceiling. So I looked for a way to orient myself: to reconstruct the world around me, from my bed, from my room, as formerly, as a child, I had done on awakening. But, though I could easily evoke where the Val-de-Grâce was located in the city, I had arrived there too fast in the ambulance and was too immobilized on my bed to place the room itself. Since sunlight never penetrated, I concluded that the window faced north, thus toward the Seine. From there, I could locate the Boulevard Saint-Michel beyond my feet, the Boulevard de Port-Royal on my left, behind the door and the corridor. At this point I could begin my imaginary walks across Paris. Go down to the hospital garden. Turn to the left. Go over to the Boulevard Saint-Michel. Proceed down it to the Seine. Walk as far as the Louvre. Go through the Tuileries, the Place de la Concorde. Go up the Boulevarde Malesherbes, pass the church of Saint-Augustin, and arrive at our apartment house. But there I stopped, before the door. I did not want to go up. I did not even dare wonder what state the apartment was in or who was living there. Not for several years had I had any news of my father or any member of our family. I did not know whether they were still alive, whether anyone had been deported. I did not know where they were. In my hospital room I had no way of looking for them or of finding out about them. I had to wait for one of them to show up. That took some time.

Another imaginary walk through Paris: I strolled through the Latin Quarter. I turned left onto the rue de L'Ecole de Médecine. I came up to the Faculté de Médecine. There I would run into some of my comrades and a few of my former professors. The welcome I received, enthusiastic or indifferent, admiring or ironic, depended on my mood and state of fatigue. I never wearied of imagining these reunions on the stage of my ceiling.

The evenings, the nights were never-ending. Reading tired me. I had to use my one free hand both to hold the book on a level with my eyes and to turn the pages. Very soon, my arm fell back. Happily, there were the newspapers. What a thrill to go through the Resistance press, just emerging from the shadows. The names still smelling of gunpowder. The meager little newspapers, badly printed, badly laid out, which smelled of ink and blackened fingers and sheets. But still they screamed fury at the Occupation; exploded with joy in the intoxication of victory; shouted, threatened, proposed, discussed, debated, criticized, planned, elaborated, legislated. And, above all, they radiated hope: hope for a new world combining justice and liberty; hope for a nation to reconstruct, body and soul, on the rubble of a shameful past. All this accorded with the sense of crumbling, of disintegration, I had suffered in June 1940 and with the necessity I had felt so strongly since of having to rebuild everything, personal life and public world. And then, every day, the papers brought news from the front. The lightning progress of the Allies. The advance of the DB. The epic of my buddies, the thundering cavalcade in which, however, I could not yet resign myself not to participate.

Fairly soon, fatigue overwhelmed me. The papers fell from my hand. My eyes grew blurred. Very gently I flowed into a sort of half-sleep. Words and letters began to wheel around me, coming together or flying apart, mocking me in those endless combinations that I endlessly mulled over, as when I was a child. Signs without significance. Pairings of syllables. Mouth games that allowed me to taste the flavor of words in unnatural alliances. The reaction of words that ceased to inform me about the private life of things, that emptied themselves of all sense, to be filled, by proximity, with hitherto unsuspected contents. I managed, thus, to split the spoken from the written. To dissociate what caressed the eyes along linear constructions of little black marks on a white page and what came to do a purring dance in the space between throat, teeth, tongue, and palate.

To start the dance, to direct it, I had the names, in capital letters, of the newspapers scattered on the bed. Written, the word *Combat* lost the violence of the spoken word, which burst like a cannon shot or a thunderclap or a punch below the belt. Rather, the paper's name evoked the velvety warmth of a just cause, the soldiers' canteens of fighting France. It savored of the countryside and of companionship. The name *Libération* had the lightness of liberty. It promised the return, in berets, of the prisoners of war. It laughed with the light murmur of a whistling breath, the rustling of pages one is leafing through. But it was the *Franc-Tireur* that, above all, fed my daydreams: because of the two letters *FR* that clove the air to announce *France*, *Franck*, or *François*. A *frisson* that flies fleetly between the teeth, that rumbles in the throat and makes the lips quiver, whistling, before stopping short, as by a sudden braking. Then began a fraternal and frustrating frenzy of words with neither head nor tail: malefactors and riff-raff fricasseeing a feast; fragments of frills, frippery, and frocks; fright fraught with defrauders and other frivolous felons; the fragile frou-frou of a frisky frump. Up to the moment when the words flew into pieces, decomposing into miserable little heaps of letters. The *F* came to plant itself on the ground, on a grave, like a Cross of Lorraine cleft with an ax. The *R* rolled, purring, streaming with a thousand fires, like the hood of a Rolls-Royce. Then the letters disarticulated themselves, got out of joint, leaving in their place a heap of nails, screws, and rods that recalled the Meccano set of my childhood. Like a heap of scrap iron in a shed to be assembled, having an appearance of logic, the tail ends of ideas flying around in our heads. I then sauntered about with my tool kit filled with pliers, hammers, screwdrivers, ready to screw in words and bolt down sentences.

This life of immobility in a hospital room scarcely favored recovery, the return to society. But the Val-de-Grâce also had its good sides. First the nurses: skilled, deft, efficient. Especially one nurse: merry, tall, sturdily built, a sort of bossy Juno, with powerful shoulders and a wide grin. It was she who bandaged me, washed me, fed me. A surprising blend of strength and gentleness, of grace and energy, who added yet again, without displeasing me this time, to my feeling of impotence.

And then there was a roommate, who changed several times during my stay. After the dead one was taken away, there appeared a great strapping bearded man, an engineer, taken on when the DB passed through Paris. He was a jolly fellow, always joking even though his leg had been torn off by an antitank shell. His wife was a dancer at the Opéra. In the hierarchy of the corps de ballet, she occupied, I believe, the rank of *sujet*. That brought us many charming visits. Every day, from one o'clock on, the room was invaded by a swarm of ballerinas. Whether student dancers or stars, they came in relays to make sure someone was always on duty. They arrived, laughing and chirping. They whirled around the room, coming to perch on the beds, rising to try out an entrechat, kissing each other, pawing about each other, fixing their makeup, trying on each other's clothes, endlessly discussing the next auditions at the Opéra. Delighted, the bearded one presided, spoiled as a pasha. As for me, I unfortunately was still too weak to profit fully from these charming apparitions.

After arriving at the Val-de-Grâce, I remained for two weeks with no news of my father or any other relative or friend. The telephone calls people made for me to various numbers from before the war produced no result. The first sign came to me from a wounded man in the DB, hospitalized on another floor. A few days earlier, he had encountered, at the Val-de-Grâce, a girl of seventeen or eighteen whose charms he described with lyricism and emotion. She said she was my cousin and was looking for me before going back to her home in the provinces. Thinking about it, I concluded that it must have been Micheline, my cousin from Lyon. But it took me some time to realize than in four years a little dumpling at the awkward age can ripen, can metamorphose into a beautiful girl.

Some days later, my Uncle André showed up, one of my father's two brothers. A tall, slim man, with a long face and an elegant manner. Freed after two years as a prisoner of war in Germany, he had gone into hiding in the provinces, along with his wife and two sons, and had just come back to Paris. Despite the ordeal he had endured, he had changed hardly at all. A bit grayer at the temples. Features a shade more gaunt. But his posture was still upright. His glance still that of a ladykiller. With Uncle André I regained a foothold in my past. I found myself in familiar territory. When, in Africa or in England, I had imagined my return to France, I knew that the landscape would be changed. That I would not occupy the same square on the board as I had before the war. That I would not find the same life with people who would be the same, as if nothing had changed. As if the war were a simple parenthesis, an interruption of the flow of time, which would afterward resume its course. But at the time I had no way of gauging the extent of the changes, or the direction they would take. I had no landmark. Uncle André provided me with my first reference system. With him I measured the discrepancy, the divergence between the life I had known before I left and the life led under the Occupation. It was a face familiar from my childhood who described for me those terrible years. Who told me how our family had gotten through this period. How they had lived, hidden, encircled, perpetually threatened with the absolute worst. A miserable life. But without calamity. Without deportation, at least for any of my close relatives. Of distant cousins, friends, however, many had disappeared. He spoke, leaning over me, while I searched his features for the secret of those who had lived for four years under the Nazi boot. And in his eyes I could read the same curiosity with respect to me.

What I was looking forward to above everything was to see my father again. But, prepared though I was for these reunions, his entrance some days later into my hospital room gave me a shock. First, because his arrival marked, in some way, the end of the running away, the return to the fold, perhaps even the hour of judgment. Also because, suddenly, his presence, alone, cast a brutal light on the absence of maman. In my memory, her death coincided so closely with my leaving for England that I had not yet fully realized that my return would not bring me back to her. Her death still had for me an unreal aspect. It was an intellectual idea, not a physical sensation. But seeing my father alone made me feel to the core of my body that it was over: I would never see her again. On his arrival, my father seemed to me very moved. For some minutes, in our joy at seeing each other again, we could not speak. He had aged a little. He stood less straight than before. His bearing revealed a certain weariness. Above all, his gaze no longer had the glint of challenge that impressed me so much as a child. Poor Papa! What years he had lived through! In a few days, he had lost a wife whom he adored, his home, and a son, of whom he knew neither his whereabouts nor what he was doing. Since then, he had lived with his mother in hiding, first in Vichy, then on the outskirts of Grenoble, constantly changing addresses. Sitting next to my bed, he was speaking feverishly, his eyes fixed on me, tense, nervous. He told about the hardships of daily life, the hunger, the risks run. He told about how he had been summoned to the police in Vichy who had strong reasons for thinking that his son had joined de Gaulle and the Free French. How he had, several times over, almost gotten himself arrested by the Germans. How, each time, he had owed his salvation only to chance. He told of his ties with the Resistance in Grenoble, for which he had become something of an expert in forged identity papers: that beats all for someone who would not consider using a second-class ticket for a trip in first class.

While he went on speaking, I felt his unease. He shifted around in his chair. He kept eyeing a pack of American cigarettes on my night table, he who bragged of having quit smoking more than fifteen years before. Suddenly he took a cigarette. It was all I could do to persuade him not to light up. At first I attributed his agitation and anxiety to his concern about the seriousness of my wounds. Or to his fear of finding me too altered, too different from what he had expected. I wanted to reassure him, to say something light, tender, as father and son so rarely do in normal times. I set my free hand on his two hands and held them prisoner for a moment. He gazed at me. Then, taking the plunge, he announced his coming marriage to the widow of his youngest brother, Pierre, who had died during the war of a sudden illness. He gave me an anxious look, as if seeking my approval. I was dumbstruck. He was right to want to marry my aunt. They got along beautifully. The two of them would make a perfect couple and have a happy life. Nevertheless, at this moment, while I was seeing my father again for the first time, while I was becoming truly conscious of my mother's absence, the news flabbergasted me. Definitely, this return had many surprises in store. Here everything had changed.

For the others who came, one by one, to visit the glorious wounded soldier, had also changed. Relatives. Cousins. Friends who filed in, examining me like an animal in the zoo, with, in their eyes, a mixture of pity, admiration, and jealousy. Of reproach, also, for having left France and thus having been spared the trials they had undergone. And then came the comrades from medical school who had gone on with their studies, and many of whom had become interns. Despite their claims, very few had actually participated in the Resistance. They went into ecstasies about Free France, singing the praises of de Gaulle and his valiant warriors. What irritated me about them was that they hailed the victory of 1944, not the choice of 1940. And, despite their enthusiasm, I felt that they took me for a fool. A fanatic. A romantic. Very endearing, to be sure, but hardly realistic; one who had preferred to go and get himself clobbered because of his idealism rather than concern himself with the health of the French and his own career.

And then one day, a new jolt to the heart when the door of my room framed a face that had made me dream so much, that for four years I had so often re-created in my imagination: the oval face of Odile. She sat on the edge of my bed. Overcome with emotion, we remained silent for a few minutes, looking at each other. Moved at finding that head, some of whose details I had forgotten and that I did not recall as being so animated and expressive. But she, too, had changed. With a beauty perhaps more affecting because it was graver, more mature, almost poignant. A better defined profile. More accentuated cheekbones. A higher forehead. A less dazzling smile. Eyes still as lively, as sparkling, but so large that they seemed to contract her face, to have moved its center of gravity. She had become engaged during the Phony War; but her fiancé, taken prisoner in 1940, was still in Germany. For four years Odile waited for him to come back so they could get married.

Like all my visitors, I looked up at her from my back. Shapes, both familiar and unknown, stood out against the rectangle of the ceiling. As she spoke, her voice, her face, her body, which I confronted with my image of her warped by memory and suggestion, seemed to me, in spite of myself, to lose a little of their magic. For her part, I felt her looking at me with more sweetness than before, but with less disturbance. More maternal, less in love. I tried to take her hand. She left it in mine a moment before abruptly withdrawing it. With a sad little smile, as if to show me that the liking she once, and perhaps still, had for me ought now to give way before her commitment to another. An odd tête-à-tête, in which each was already beginning to look at the other as a memory; in which, with emotion and tenderness, we seemed to be leafing through an album of photographs already yellowed with time. Each of us spoke warily, for fear of wounding the other. As if, by a tacit accord, we should avoid destroying what, for each, must remain an unforgettable youthful love. She kissed me before leaving. And when she was gone, I was touched to anguish by the sense of a world that would never return.

Everything was foggy. Eerything was blurred. There were colors, iridescent flashes, but no shapes. I no longer knew how to use a microscope. Instead of adjusting the eyepiece to my sight, I squinted as if taking aim. The only things I could pick out were pink or mauve plaques. But no patterns, no structures. Finally, by constantly moving the slide, by fiddling with the focusing knob, I succeeded in detecting some small rectangles. They were cells, each with a darker mass in the center: the nucleus. A certain regularity. So what kind of tissue was this? Shit! Four years of war to end up here! Get my neck broken to be back again in a histology exam like a kid and not remember a thing about the subject!

This was one of my first outings from the Val-de-Grâce. An afternoon leave. Limping, leaning on a cane, my arm in a sling, I dragged myself with my father to the medical school to try my hand at the second-year examination that I had been unable to complete in June 1940. Histology and biochemistry. Needless to say that since then I had forgotten everything. Needless to say also that there was no question of my studying in the hospital for these exams. No surprise, then, to find myself an ignoramus, stupid before my microscope. With his pink scalp from which the rare white hair emerged, the examiner began to hover around me. Before my departure, Professor C. was thought to be a bastard. Since there was little reason to suppose he had improved in the past four years, I saw myself off to a bad start. He came up. "So?" he began. I took the plunge. "It's the liver." "No," he answered dryly, "it's the kidney. That gives you a grade of five out of ten." Wrong, but I passed! For a moment, I remained there with my mouth open, staring at him. He had already gone on to another student. Clearly, it was to my uniform and decorations that I owed this grade. Which seemed to me perfectly normal.

Things went less brilliantly in biochemistry. Of the three examiners, one was blind. A laboratory explosion some years back. It was my misfortune to get this blind man who would be swayed by neither the uniform nor the medals. He asked me about the metabolism of succinic acid. I sat speechless, unable to think. Succinic acid, of course, reminded me of something. I remembered having known its properties, its reactive capacities, its function. But there, before the dark and empty glasses of Professor J., *succinic* was only a word. A word that I sucked on. *Success. Succeed. Succumb.* I got one out of a possible ten, which was generous. Happily, another professor whom I had known before the war noticed the fix I was in, and had me triumphantly take a makeup exam. And in the evening, when I returned to the Val-de-Grâce, I had at last managed to pass my second year of medical school!

To go, to come. To arrive, to leave. To enter, to exit, to stay. To be born, to die. To ascend, to descend, or to fall. A series of intransitive verbs that Monsieur G., the ninth-grade literature teacher, had us chant over and over. Years later, it was the hospital that had me harping on these verbs once again. I went to the hospital for months on end. I came out of the hospital. I entered the hospital for a new operation. I stayed in the hospital. I left to return there a few days later. It was never-ending. Nearly a year went this way. On 8 May 1945, I was there once again to have a piece of shrapnel removed from my thigh. Green with rage at being unable, once again, to participate in that explosion of joy that marked the signing of the armistice, the end of the war in Europe. How to put my life back together under these conditions? How not feel out of step with the world, with other people? Had I not been so badly wounded, perhaps I would have been able, despite the war years, the desert years, to take up where I had left off. Perhaps I could have slipped back into my former existence, into my chosen profession, as into some old clothes one finds, one day, still in good condition.

I arrived ahead of time at the Office of the Assistance Publique, the central administration of Paris hospitals. I needed time to consider how to present my case. Some days earlier, I had abruptly decided to stake my future as a doctor on a bet. Everything conspired to discourage me from following this route. First, because of the aftereffects of my wounds, it was out of the question for me to become a surgeon: this profession admitted of no handicap. Then, I felt less drawn to·a medicine that lacked the sporting and dramatic aspect I loved about surgery. Finally, I was fuming at the idea of finding myself with students much younger than I and of seeing my contemporaries as brilliant hospital interns while I was not even an *externe*.* Hence, my resolution to practice medicine only if I could rapidly become an intern. If not, that was it: I would finish my studies as quickly as possible to get the diploma, and then do something else. So I had to become an intern right away without being an *externe*. Which violated the sacrosanct law of the Assistance Publique. Which, in the memory of any medical student, had never occurred.

That was what I went to plead at the Assistance Publique. I had talked about it with my former supervisor Thérèse Bertrand-Fontaine, the first woman to have been granted the title of Head of Department in the Paris hospitals. In 1940, I had worked for nearly six months at her clinic in the Lariboisière Hospital, learning there the little medicine I knew. In 1944, on my return, she alone of all my former supervisors demonstrated a warm friendliness. When I had told her of my hope of presenting myself directly for the competitive exam for the internship, she'd had a skeptical smile. But she had encouraged me to launch an assault on the Assistance Publique.

I was received with great cordiality. A tall, thin man: a sober suit, slicked-down hair, steel-rimmed glasses. He asked with interest about my peregrinations, my campaigns; sympathized about my wounds; recounted the horrors of the Occupation; testified to his admiration for Free France, to his veneration of Leclerc. A bit too good to be true! After this preamble, I began to plead my case. A first competition for the externship, set for

* In order to work in a hospital, a medical student had to take two successive competitive exams: *externat*, during the second year of medical studies; and *internat*, in the fourth or fifth year. Before taking the *internat*, one had to have been an *externe* for two years.

December 1939, canceled because of the war. The delay for the medical students who had been away fighting for years or who had been prisoners. The injustice of a situation that favored those who had remained in France. Our hope of being able to present ourselves directly for the internship exam without having been *externes*. Immediately, he stiffened. I thought he was going to have an epileptic fit. Smiling up to this point, his face became set. As if I had uttered an obscenity. As if I had personally insulted him. I set out my arguments. What I was asking was not a favor but simply permission to take a competitive examination. To be graded in the usual way. Either one was accepted or one failed. But my interlocutor was no longer listening. For a moment I continued to explain, to think up new arguments. Then, suddenly, I had had enough. I walked out, slamming the door behind me.

I could no longer stand the military surgeons at the Val-de-Grâce. All winter long, they had been probing my wounds and trying to remove bits of shrapnel with pincers, without anesthesia. In the spring, I still had a suppurating scar. This time, I was operated on by civilian surgeons at the Léopold-Bellan Hospital. The final operation. After which I left the hospital for good. It was the time of the great return—the return of those who had joined the army; the return of prisoners of war.

The return, above all, of the deportees, of the people coming back from the death camps. Skeletons of withered skin showing through their striped rags. With tales of horror. Worse than anyone might have feared, than anything one might have imagined. The revelation of a world where, at every moment, the impossible became possible. A sort of reverse nightmare where waking up, far from relieving one of anguish, showed up misfortune in all its sharpness, plunging one back into hell. And hell was not merely hunger, fleas, typhus, wounds, blows, madness, corpses, the horrible smell of the smoke rising from the crematoria. It was, even more, the organization of debasement; the machine for humiliating, for breaking the human spirit, for forcing it to hate itself. A task that, up to this point, humanity had not yet invented. For torture formerly had the purpose of extracting confessions, of repressing political deviance or religious heresy. The ultimate goal, here, was to degrade. To reduce the prisoners to the condition of animals. To make them lose, in their own eyes, their soul, their dignity as men and women. How otherwise are we to understand the stories of soup spilled to force the starving to lick the ground? Or of political prisoners subjected to the authority of common-law criminals, pimps, thieves, and prostitutes? Or shut up with madmen? Or used as guinea pigs for the "experiments" of demented physicians? All these stories seemed out of time and place, for the atrocities to which history has accustomed us had taken place far away and long ago. They referred to the anonymous people of the past, to people without faces, to the others, the slaves of the pharaohs or the people conquered by Rome or Attila. The Nazi horrors, on the contrary, were experienced here, now, at home. They had touched our neighbors. Our friends, our relatives. Each one of us could have been hurled into a train to arrive, in a night pierced by flashlights, at a railway platform shaken by the howling of dogs and the blows of the SS. The victims were the same as ourselves.

One night we had been on a binge, Yves C. and I, and we wound up in a club in Montmartre, Le Blue Moon. Semidarkness. Soft music. Tobacco smoke. A few customers, some sitting at tables or dancing on the dance floor. At the bar, surrounded by girls, a tall, fairly young man of an alarming gauntness, drowning in his clothes, his eyes sunk deep in their sockets: the very picture of the death-camp survivor. When he noticed me, he began to stare. The girls around him stopped laughing. He pushed them away. Then he came over to me. "You don't recognize me?" I hesitated for an instant. But behind the little mustache, I detected the harelip that back in the lycée made Jean B. look as if he were always smiling. Which annoyed the teachers. He was younger than I, but we had acted together in an amateur production of Giraudoux's *Tessa* at the home of friends. Jean stared at Yves and me by turns as though to see what impression he was making on us. Then he pulled up his sleeve to show the number tattooed on his arm.

The three of us sat down at a table. A fat blonde came to sit with Jean, who touched her distractedly. The band had stopped. After a silence, I asked Jean how he had been arrested. Whether he had been in the Resistance. "No, I was too scared," he muttered quickly. For a long time, he had not wanted to believe in the horror, in the Gestapo. He was living with his family. He avoided going out in public, breaking the regulations. He had even worn the yellow star. Sometimes a pal would vanish, but he asked no questions. Until the day when, near his home, in the street, he had see two bruisers in black hats punching a man in the face while they shoved him into a black Citroën. People passed by. Nobody said a thing. Some weeks later, in a café, Jean had seen the two men in black hats come in. They were looking for someone. Then they left. And one morning at five o'clock, they came for Jean and his parents, to take them away. "We should have left before," said Jean with a faraway look in his eyes. "But my father was sick. And we didn't have any money." Jean was ashamed of this poverty, of his family. For a long time he had thought that everyone was poor. That that was life. Until he entered the lycée, where he had seen others, where he could compare. Since his arrest, he had had no word of his parents. They must have died in some camp, in some gas chamber. As for him, he had been taken to the Gestapo. He had been arrested as a Jew. But that was not enough. In addition, they wanted to make him confess that he was working for the Resistance. He had been beaten. He had been tortured. He had been made to stand for four solid days. Without eating. Without drinking. Forbidden to lean on a wall.

The band had started up again. On the dance floor, two couples were moving about in a frenzy. The fat blonde had discreetly left our table, to be replaced immediately by a pretty little brunette with a plunging neckline, who came to sit next to Jean. He was telling us about Dachau. The blows. The dysentery. The lice. The unspeakable soup. The fear. The never-ending humiliation. People died every day. The corpses piled up in a corner of the camp. The horrible smell. The prisoner-guards, always on your back. One day, with three other inmates who did not stand aside quickly enough for some passing SS men, they had to dig their own graves. Two hours of turning over earth with shovels. After which they were made to stand at attention for four hours beside the ditches. And then, for no reason, they were ordered to fill up the holes. They were sent back to work. They never learned why they were not executed.

Once again, the band stopped. The room had become dark. Two searchlights were sweeping the dance floor. Then the band started up again. Four dancers appeared, one of them the fat blonde who had just left the table. Adorned with feathers and paste jewelry, shoulders and legs bare, one breast exposed, each moved in rhythm eagerly and painstakingly. Two steps to the right, a kick in the air; two steps to the left, a kick in the air. Passing by our table, the fat blonde smiled seductively at Jean, who did not respond. On his return to Paris, Jean had spent some days in the hospital. After leaving, he had not known where to go. One evening, he had turned up at Le Blue Moon. Since then, he came back every evening, spending most of his nights with the fat blonde. He was not doing anything. He had no desire to do anything. I would have liked to help him, but he would not hear of it.

A month later, I returned to Le Blue Moon to find out how Jean was doing. He had vanished, leaving no address.

Once again, I had to make a trip back to the Val-de-Grâce for a follow-up visit. Afterward, I went for an aimless walk. Walking in a Paris of snow and mud. Haunted by ghosts. By the little jockey from Saint-Jean-de-Luz and by Vincent D., the soldier of Bir-Hacheim and of Jebel Matleb. By Roger D. and his brother Jean, the one murdered in Chad, the other in Algiers. By Jacques S., the brilliant flying commander shot down in flames, escaping and later disappearing into the sea; and by Jean B., the little Jew deported to Dachau who would never recover. The long list of the dead, in combat or in crematoria. Of twenty comrades, officers in the Marching Company in 1940 at Dakar, only three returned. In the First World War, death took men at random, without any relation to their merits, their value. In the Second World War, death was more selective. Those who had fallen were the best. The most resolute. The volunteers for the Resistance or Free France.

In these black and icy streets, I was cold. I was going through one of the oldest, most depressing areas of the city. The one where had been built the Salpêtrière and the Cochin hospitals, the Sainte-Anne insane asylum, and the Santé prison: the home of the ill, the mad, the condemned. I was walking in a world of the derelict, in an area of poverty. And I was cold. Along walls oozing dampness and misfortune, I saw the long cohort of my friends dead or disappeared, one day, at the age of twenty. For freedom. And I was cold. And, little by little, going down toward the Seine, I felt rising, from the deepest part of myself, the irresistible desire to have a child.

Since I was not to be a hospital intern or *externe* but wanted to finish my medical studies, I needed to complete a series of rotations in obstetrics, oto-rhino-laryngology, pediatrics, psychiatry, and so forth. Each one would last two to three months. No sooner had I begun one of these rotations with Professor T. than a friend from the Second DB, posted to the cabinet of the minister of war, ordered Yves C. and me to go on a mission in Germany and Austria. The war had just ended. By sending us to gather information about the Nazi hold-outs in the Tyrol, André J. also wanted to give us a holiday. So I went to find Professor T. to ask him to give me the credit, after a week, for a rotation that was supposed to last for twelve. I wanted to see him alone, which was difficult, for T. did not make a move without his staff interns, his band of nurses, and his battalion of trainees. Nevertheless, I managed to corner him in his office and explain my request. He started. "It's out of the question to give you your rotation after one week," he said with a derisive laugh, and showed me the door. I then exhibited my orders. At the sight of the document, stamped with the tricolor and signed by the minister, his expression softened. Incredulous at first, he examined the document in detail. Tall, thin, his eyes tending to pop out of his head, Professor T. was said to have exhibited little opposition to the occupying forces during those years. In a gentler tone, he asked, "And what are you going to do there?" "I'm sorry, monsieur, but I can't tell you that." Dumbstruck, he looked at me, suddenly full of respect. He signed my trainee certificate. Then, in a tone that was meant to be full of complicity, he declared, "Go to it, monsieur. France calls you!"

A few days later, Yves and I established our general quarters at Saint-Anton in Austria to find out about the Nazis hidden in the area. One afternoon, I went to a barber for a shave. I picked up a newspaper. That very morning, a new type of bomb had exploded in Japan. It had, with one blow, razed the city of Hiroshima.

A visit from P. He wanted to see me urgently. We had not seen each other since Chad, where we were posted together in the Twelfth Company. A second lieutenant in the reserves, he had a small factory in one of the suburbs of Paris. In Africa, P. was obsessed by the memory of his wife. She was all he talked about. Sometimes we would see him go off, alone, in the heat of the night, his face contorted. He would walk around for hours on end muttering, "She's cheating on me! I'm sure she's cheating on me! When I get back, I'll kill them both!" He had been wounded in Libya and, since then, I had heard nothing of him.

When he arrived at my place, P. was smiling, relaxed. He had recovered his factory and his wife. Life was full of promise. Except for one thing. He wanted a child, but had just learned that he was sterile. So he had chosen me to make a child with his wife! Astounded at first, I came to feel both amused and flattered.

Worried, too, for I could not forget his crises of depression and jealousy in Chad. In any case, it was out the question to comply with his request. If one day I was to have children, it would be to care for them myself. But it did no good to say this. It did no good to defend myself, to explain to P. that I could not do something like this to a friend, to the wife of a friend, to a war buddy. He didn't want to hear about it. I must come and dine with him and his wife "at least to make her acquaintance." She was beautiful: small, slender, with a mass of blond shoulder-length hair, sparkling dark eyes. She must have been aware of the situation, for she seemed as embarrassed as I. Only P. smiled, perfectly at ease. As soon as dinner was over I fled. With regret at not having met, outside of conjugal blessing, such an attractive woman.

The postwar world. Postwar France. We had discussed them endlessly in the sands of the desert. Once freedom returned, how could social justice be enhanced? Had I seriously believed that victory would be accompanied by a revolution? Or that the same spirit that was to crush Nazism would build a country so new that power would no longer be in the hands of a bourgeoisie dominated by money? Perhaps I had thought this way in Africa, far from all reality. But when I was back on my feet after being wounded, the question no longer arose. On leaving the hospital, when I began to get around a bit in the city, the enthusiasm of the liberation of Paris, the calmer one of the signing of the armistice, had died down. Life went on. The life of 1939, after the intermission. Most men and women had looked at the drama unfolding as at a passing storm. Their main concern, what in their eyes gave spice to life, was to hold on to their goods, to increase them, and to enjoy in peace the benefits they brought. For these people, the nation's misfortunes were confounded with the attacks on their pocketbooks. For a long time they had looked on de Gaulle as a traitor, not to the nation but to their interests. Until the day he had appeared as the defender of public order against the communist threat.

After getting out of the hospital, I lived in a little room on the premises of my father's office, near the Parc Monceau. No modern conveniences, but some independence. In the morning I had breakfast in a nearby café with office workers. The rite of croissant and café crème was a tangible sign of my return to Paris, along with steak and French fries, walks along the Seine, the bookstalls, the bus platforms, the cries of newspaper vendors. The owner of the café had been a noncommissioned officer. Wounded in the Great War, he had obtained a tobacconist's license. A great husky fellow, very sanguine, a loudmouth, with something of a Toulouse accent, he was ensconced behind his counter, bossing his café like an infantry squad. As he had seen me several times in uniform, he had a soft spot for me and called me "lieutenant" or "doc." In the neighborhood he was considered a major figure in the black market. He had gone effortlessly from Pétain to de Gaulle. He had even been deeply affected by the general's relinquishing power in 1946 and withdrawing from politics. "Something should be done, lieutenant," he said. "We can't just let him go like that. All those politicians, they'll go back to their crap. They'll soon have us back in the shit. We should do something!"

This was also the opinion of some of my buddies from Free France. Many took poorly to the return of normality. To a regular life. To work. To routine. For during those four years of war, they had surpassed themselves. They had risked their lives. Lived an epic. Written history. It was hard to give up, just like that. Many were trying to continue the game in one way or another. With the war in Indochina, for example. The same was true of those of the Resistance who had lived in hiding, outside the law, without a home, without even identity, constantly moving, mystifying the police and the Gestapo, gambling with their lives and with those of others. They did not easily accept taking up old habits, returning to the ranks, living out in the open in a world they found cramped and shriveled. The clandestine had become their life; the secret services, their drug. For them, the state of peace was a state of withdrawal. Several times I was approached by former comrades wanting "to set up networks" so as to "maintain the spirit of Free France and the Resistance." But in no way did I feel addicted to that drug. I had had my fix!

Among my old classmates at the lycée or medical school whom I then ran in to, several had, in the Resistance, joined the Communist Party. Some of them chose me as a target for their proselytizing. I had never been tempted to join a political party. Except perhaps the Socialist Party at the time of the Popular Front. But very quickly the refusal of Blum's government in 1937 to send arms to the Spanish republicans and, in 1940, the voting of full powers to Pétain by most of the socialist members of Parliament, had killed my wish to join their party. For a long time I had only a bookish knowledge of communism: epic in the case of André Malraux, flattering with Maxim Gorky, irritating with Victor Serge, interpretive with Arthur Koestler. At the Liberation, the Communist Party glittered. Anticommunism and anti-Sovietism were no longer in fashion. Stalinist communism appeared as both the martyr and the hero of the war. It was earning interest from all the struggles, all the sacrifices, all the suffering, from all the oppression experienced over the years by the proletariat of the whole world: czarist deportations, Spartacus uprisings, strikes of all kinds and in all countries, the struggle of republican Spain, and, above all, victims of the war, the French communists shot, and the millions of the Soviet dead. Moreover, the "party of the people shot" had also become the party of power. The weight of the Red Army, principal conqueror of Nazi Germany, endowed communism with an enormous power of attraction. Once Hitler had invaded the Soviet Union, the French Communist Party had played one of the leading roles in the Resistance. If, in 1940, I had had to remain in France instead of leaving for England, perhaps I would have become a communist. At the request of my friends, I attended several meetings of communist students. Very quickly, I discovered that I was not on the same wavelength as they. Worse, I had no desire to get on it. This was in the fall of 1946, the time of the first elections for the National Assembly. There was a great deal of discussion at student meetings about how to achieve socialism along routes other than those taken by the Soviet communists. One evening was devoted to a recently published article by a member of the Central Committee who tried to prove that "true democracy" not only did not exclude but entailed the "dictatorship of the proletariat"; that the two ideas, far from being incompatible, complemented each other. Everything about these meetings grated on me: the words used, the meanings they were given, the arguments from authority, the constant references to the sacred Marxist texts, the attitude of the speakers, their devotion, and, above all, their certainty of being right, of being in possession of the truth, a truth both political and moral. There prevailed an atmosphere of mysticism, an odor of religion. Abruptly, I had the feeling that joining the Communist Party would be like converting to Christianity or Islam. My friends sent me back to my "life of the little bourgeois." I felt neither shame nor guilt.

While gradually regaining the use of my legs, I was able to take up again and lengthen my walks across Paris. Mainly, I was looking for the joyous, triumphant neighborhoods. The tourist areas: broad avenues lined with trees, gardens bursting with light, monuments glorifying the past. Despite the threats of destruction, the city itself had changed very little. One could not, however, look at it with the eyes of an earlier time. After the massacres, the tortures, after the millions of the dead, of skeletons piled up in mass graves, the world could no longer be the same. Those men, those women who had survived on the fields of battle, in clandestine struggles, in the death camps, all those who had spent years in horror and anguish, what would they now dream of? What could the people of my generation believe in, if they were neither religious nor communist? Their youth had been stolen from them; their friends killed; their hopes, their enthusiasm dashed. What meaning, what substance could they now give to words like *honor*, *truth*, *justice*, and even *nation*. Of these grand words, only *freedom* had held up. It had even taken on meaning, having been shouted in the silence of dungeons and whispered in the din of battle. It was for freedom that people had left home, family, country; that people had fought; that people had been imprisoned and tortured. No longer could we allow just anyone to do whatever they wanted with freedom. But how to reconstruct a nation no longer at the mercy of a beast or a madman? In the Nazi tempest, all the nations of Europe had yielded at the first gust. All except Great Britain. Saved by being an island—And yet then only through the alliance of the United States and the Soviet Union! But if, one day, one of these giants developed an appetite, how could we resist? How could we stand against pressure: military, economic, political, or cultural? I saw only one answer, an answer both simple and difficult. Simple to imagine because, quite obviously, the solution was Europe: a United States of Europe, the USE, the only way of preventing aggression, the temptation to madness, from within as well as without. But this was a difficult solution to put into practice so soon after the war. With the dead just buried, the deportees just returned, the idea of Europe, the plan of union with Germany was scandalous. I myself would not have wanted to visit Germany for anything in the world, for however short a time. But what passion and hatred now denied, reason would one day be forced to create. The plan of a federated Europe was, I thought, the one way to put an end to those wars that left whole nations worn out, bled white. It was also the way to give the lost generations a new reason for hope and enthusiasm. A great cause to defend. So if I had been de Gaulle, it is to this plan that I would have devoted myself. But I was not de Gaulle!

Man exudes a plan. Sweats design. Stinks of intention. Does not tolerate contingency. Nor admit that the evolution of the species occurred by chance, or that human history obeys no secret law. Some want this history to be governed by the awareness of a goal; others, by a blind fatality. But whether men do or do not believe that they are conscious of the goal, whether they consider themselves free agents or compelled to take up tasks whose ultimate reason they do not know, all men want to see history flow inexorably in the same direction. What man seeks, to the point of anguish, in his gods, in his art, in his science, is meaning. He cannot bear the void. He pours meaning on events like salt on his food. He denies that life bounces along at random, at the mercy of events, in sound and in fury. He wants it always to be directed, aimed toward a goal, like an arrow.

The first bell I rang was that of the office of the competitive examinations at the Assistance Publique. For months, I rang many another. If I were not to become a doctor, I had to have another destiny. I did not know what to do. I wanted to do everything. Not the slightest sense of vocation. A total absence of spirit and inclination. Ready for everything and for nothing. Capable of everything and of nothing. Wandering. Lost. Pulling in opposite directions. Full of arrogance and shyness. Wanting to attract attention. Not daring to show myself. Seeking what I could not have. Refusing what was offered me. Trying here. Trying there. Looking in all directions. Journalist! Writing articles. Selling them as a freelancer. In and out of newspaper offices. One article accepted. Ten refused. Film actor! Why not? A questionable attempt. Not really a failure. Not conclusive, either. Better not. Politics! The ministerial offices. After all. Perhaps. An interview with Louis J., secretary-general of the government. He tries to discourage me. He succeeds. Fortunately. I have neither a feel for the thing nor the necessary personal qualities. The new National School of Administration. A high-level civil servant! That's the thing to be! Didier P., a war buddy, convinces me that my future lies in that direction. A special examination for those in the Resistance and the Free French and for prisoners of war. Tailored for us. I buy a law book. On the third page, it falls from my hands. Farewell to administration. A holiday with a group of friends. One of them, the president of a major bank, wants at any cost to have me on board with him in finance. No taste for business. No reason to get myself trapped. No to the bank!

Less than brilliant, all that. Just time wasted. Desires repressed.

Ambitions thwarted. In the final analysis, nothing serious to get my teeth into. Nothing to furnish my insomnia, to feed my dreams of the future. Nothing to feed me, either! Over twenty-six years old, I was living on my savings from the lieutenant's salary accumulated during my stay in the hospital. I was not earning a cent. I had no profession.

And then my medical studies, the rotations, the examinations to complete. As quickly as possible. Doing two years in one, to make up for the time lost in the war. A good way not to learn anything! And one day, at last, all studies completed, all obstacles overcome, there remained only a thesis to submit before becoming a doctor of medicine. Given my ignorance and lack of enthusiasm for the practice of medicine, I needed a tangential topic, on the fringe of clinical practice. Opportunity came in the form of a strange organization more or less derived from the Second DB.

In 1938 when I first went into a hospital ward, a miracle word, *sulfamide*, was just beginning to circulate. To universal wonder, this drug cured contagious diseases that until then had been thought fatal. It was the start of a revolution that was to remake medicine. But at that very moment, when I entered medical school, no one (and I mean no one) had even a ghost of a suspicion of what was about to happen. For our generation as for the ones before us, the medicine we were starting to learn seemed necessarily the one that would serve us all our lives without changing significantly. Toward the end of the war, another miracle word was circulated in the Allied armies: *penicillin*. Discovered and isolated by the English, this compound was mass-produced by American industry. The French were not in on it; to treat their sick, they had to be content with whatever amount of the drug they were given from abroad. In 1945, at the initiative of a doctor from the DB, Major B., the army created a center to study and eventually to produce penicillin. A garage on the rue Alexandre-Cabanel in the Fifteenth Arrondissement: laboratories hastily set up; a few doctors; a few biologists; a few engineers; a few plumbers and electricians. The organization was baptized the National Penicillin Center.

The manufacture of penicillin was then no small matter. It had taken years for the best English biochemists to purify the active agent, and months for the formidable resources of American industry to mass-produce it. The precious strains of *Penicillium* were their private preserve. The procedures remained secret. Compared with the Anglo-Saxon professionals, the amateurs of the rue Alexandre-Cabanel were not heavyweights. For months on end, they had to confine themselves to retrieving what had already been used; going to hospitals to collect the urine of patients treated with penicillin; concentrating the urine; extracting from this juice the active agent. All this to obtain a few grams of the precious powder each month. Do-it-yourself work.

At the end of the war, antibiotics constituted a new, and still relatively undefined, field. Just what I needed. There I was not out of place. My own ignorance merged into universal ignorance. With a little penicillin, I said to myself, one ought to be able to get results with little expense of time and money, and thus make quick work of one's thesis research. So I looked up Major B., a doctor I had known in the DB. Short, slight, black-haired, he was both ambitious and headstrong, sometimes friendly, sometimes insufferable, going without warning from anger to joking. In matters of research and industry, he was not overburdened by scruples. On a visit to a competitor, he never failed to take the panoply of pirating tools that enabled him surreptitiously to take samples of *Penicillium* strains. He gave me a good welcome, so pleased was he to contribute, like a real medical boss, to the production of a thesis. For a topic, he proposed studying not penicillin but another antibiotic called *tyrothricin*.

Only a minor antibiotic, tyrothricin was far from stimulating the same interest as penicillin. Too toxic to be injected, it could nevertheless be used for certain localized infections. This product had been isolated some years earlier by René Dubos, a Frenchman working at the Rockefeller Institute in New York City. The thesis project consisted of making a little tyrothricin and then studying its therapeutic properties in certain diseases. At the request of Major B., Dubos had generously sent along strains of the bacteria producing tyrothricin. So it was a matter of cultivating these bacteria and then of treating them, following to the letter the recipes published by Dubos. Of laboratory work I had neither knowledge nor experience. Only willingness. And a desire to reach my goals rapidly. For several weeks I was at sea. My culture medium remained hopelessly cloudy. My filters tore. The bacteria refused to multiply. The cultures became contaminated. Nothing went the way I wanted it to. The image I had of myself was "Charlie Chaplin goes to the lab." My motto: "Quick and dirty."

And then gradually things fell into place. The medium became clear. The bacteria started to multiply. Putting them through a centrifuge, I obtained a yellowish-pink paste with a disgusting smell. From it, with repeated lavages and extractions, I drew out a whitish powder. And this powder was found to kill certain pathogenic microbes: staphylococci, streptococci. Just as Dubos had written. I then set about going to hospitals with my powder, trying to persuade doctors I knew to use it to treat certain maladies. Several of them complied. To my surprise, my powder had good results in several local infections: boils, anthrax, throat ailments, and the like. In a few weeks, I was able to collect enough data to write a thesis. Not a very original thesis, since I had confined myself to replicating American work; but a thesis at least equal in quality to most of the ones submitted to the medical school. And, one fine afternoon in the spring of 1947, dressed in the black gown of the medical school, I, along with four "colleagues," pronounced the Hippocratic Oath. Out on the sidewalk after leaving school, I felt both happy and sad. Happy because I had finally gotten my diploma; I was a doctor of medicine; I had the right to set up shop, put a plaque on my door, wait for a customer. Sad, too, because this diploma, which should have brought me a profession, opening up a new life, was perfectly useless to me. I could not do anything with it. I was not in a position to practice medicine in a decent way. I did not know what to do with myself.

My time at the Cabanel Center, however, had taught me a

great deal. First, it had represented a new experience. Up to then, my social life had gone on among groups of students or soldiers, bossed more or less from outside, directed by teachers or superiors. For the first time, at the center, I had lived in a sort of business. A world where relations between individuals were articulated differently: with occasional conflicts of interest, with struggles over a job or a salary. I had a sense of having entered a world of adults where the place I occupied was still ill-defined, still insecure. But this community lacked a soul. There was no coordination, no common effort. A heterogeneous mixture of military men and civilians under contract. All going every which way. Major B. had neither the knowledge nor the enthusiasm nor the presence needed to animate this group. To turn them into a team.

And there, too, I had had the chance to observe professions that I knew little about. Professions on the fringe of medicine. Offstage, in the wings of the clinic. A whole world that busied itself with finding medications, fabricating them, publicizing them, selling them. Walks of life that had little attraction for me at the time. But who knows? Someday, perhaps.

Finally, and most important, I had gained an inkling of what research could be. During the research for my thesis, I had of course merely reproduced experiments, replicated ideas, redone movements that others had done before me. Nevertheless, here and there, I introduced a few variations, took a few liberties. Oh, very minor ones, to be sure. Simplifying a technique. Shortening a purification. Once I had taken soil from the square in the Place Alexandre-Cabanel looking for germs capable of producing antibiotics. I had isolated one that seemed active. I had tried to characterize it. For some days, I had thought I had achieved an unpublished observation. Made a discovery. A find. Possibly important. Up to the moment I noticed that the effect was a matter of contamination. By a well-known microbe, described long ago, which was hanging around the laboratory. I had fallen from the heights. Because, for some days, I had believed in it. I had seen myself launched into research. Into discovery. And, above all, I had grasped the process. I had tasted the pleasure. The whole attitude reminded me of my childhood. The exploration of the world. The game. The ideas that cross the mind and that one immediately tries out. Just like that. To see. One in ten succeeds. One in a hundred. Enough to give the impression of success, a sense of power. Like the child who for the first time manages to place one block on top of another. And who, amazed at his discovery, starts over, again and again. As though to convince himself that this is how the world functions. Unfortunately, in the summer of 1947, I had also made another discovery. I had gone with B. to a congress of microbiologists in Copenhagen. There, listening to papers of all kinds, I had been able to estimate my incapacity. I learned that research was not for me. That I could never play at that game. I was too ignorant. And not just too ignorant, but also too old to learn. Too old to hope to acquire that combination of knowledge, experience, and flair without which one is wasting one's time. Without which one has no chance of winning the game. Without which one remains a second-rater.

After my thesis, the center at the rue Alexandre-Cabanel was in a position to produce and to sell tyrothricin. An opportunity that Major B. could not let slip. He proposed that I stay on at the center under contract to work on the manufacture and commercialization of tyrothricin. I hesitated. I really did not care for the center. I had a few acquaintances there, no friends. I did not see where this world of amateurs would lead. "Quick and dirty" was good enough for a thesis, but not for a life. And then to continue the same work was to court boredom. Nevertheless, I accepted B.'s proposal. Why? I don't really know. Perhaps for lack of anything else to do. Perhaps out of fear of wandering a long time in pursuit of a profession. Or because of the lure of a contract that, although modest, gave me a social reason: the sense of being able, finally, to earn my living. Perhaps also in the hope of familiarizing myself with a biological industry in gestation. Lacking the education needed for pure research, I might find a place in industry. Even in industrial research, less confining, less demanding than pure research.

And then began a period of interminable boredom. The job involved working with engineers, to start up production, to develop it, to adapt it to growing demands. With pharmacists, making up aqueous solutions, ointments, gargles. There were a few problems. A few complications. Nothing serious. And, in the final analysis, tyrothricin was the only medication that the National Penicillin Center ever managed to produce and put on the market. But I did not take to the industrial game. Perhaps because the group lacked fervor, passion. Perhaps because tyrothricin was not a very interesting product. Life turned quickly to routine. I spent the best part of my time reading. Scientific books like *Evolution: The Modern Synthesis* by Julian Huxley, *Chemical Embryology* by Jean Brachet, *What Is Life?* by Erwin Schrödinger. And also novels: Camus, Proust, Kafka, Dostoyevsky, Faulkner. One day, Major B. asked me to travel down to the gunpowder factory of Morcenx, in the Landes area of southwest France. My job was to oversee the manufacture of penicillin, the primary aim of the Cabanel Center. The factory, of military origin, was attached to the "Office of Explosives," itself under the authority of the minister of armaments, then the communist Charles Tillon. With the return of peace, the makers of gunpowder wondered what to do with their factories, what to "reconvert" them to. In the minister's entourage, the idea had arisen of making penicillin in them. The idea was simple. It ran like a syllogism: the Americans produced their penicillin in huge fermentation vats; *now*, there existed huge vats in certain gunpowder factories; *therefore*, in these factories penicillin could, even should, be made. The idea was seductive. It had symbolic value; from war to peace; from death to life; the weapon of health. The problem was that the idea was impracticable. It took about ten minutes to see this, as I did one morning, getting off the night train to Morcenx. The engineers took me to look at the vats that had formerly been used for God knows what. Nothing to compare with what was found in American industry, with its stainless steel vats that could be sterilized under steam pressure. Not a prayer of ever cultivating such a delicate, demanding creature as the *Penicillium* mold in these old sheet-metal dinosaurs. Major B. must have been aware of this. So what had he sent me to do here? When, two days later, I returned to Paris and asked him, he looked at me with a twinkle in his eye and referred me to the member of Tillon's staff who had dreamed up the operation. A strange character, this gun-powder engineer, with a simple-minded look about him. Short, blond, a bad case of strabismus behind thick glasses, this graduate of the Polytechnique was also a member of the Communist Party. A deadly combination! He listened to me with a wide grin, then declared that he did not for a second believe it impossible to produce penicillin in Morcenx. It was purely a question of work and cleverness. I came away utterly disgusted, convinced that the motives for the Morcenx operation were not industrial but political.

At that moment I should have quit the center, slamming the door behind me. But I stayed on. Still out of fear of the void. Out of fear of finding myself without a profession, with nothing to do, not even knowing what to aim for. For six months, I went two or three times a month to spend two days in Morcenx, vaguely listening to the engineers tell me their troubles, their inability to cultivate the admirable fungus. Not one gram of pen-icillin was ever manufactured in Morcenx! On returning to Paris, I did a few experiments at the laboratory. Without believing in them. Six months of night trains, hotel rooms, fatigue, boredom, sadness. All for nothing. What a waste of time! What a mess! What loneliness!

A period of dejection, of bitterness. A period with no points of reference. In which images get entangled and memories overlap. In which this mixture of dreams and events, of scenery and faces, from which a life is woven, is reduced to a little mass without form or color. I still walked with some difficulty, with spasms of violent pain. In the evening, I would return to the cold of my room, among those empty desks where my father came mainly to see me, always warm, always worried about how I was doing. I had seen Odile again two or three times. Then her fiancé had come home from Germany with the prisoners of war. They were married. That enormous mass of dreams that I had so long accumulated about her was now without an object, floating like a balloon cast adrift. Everything that formerly had occupied me seemed to have ebbed away. I went from apathy to exaltation. Sometimes I saw myself as the most idiotic person of my generation, incapable of doing anything, behaving with women like a child staring through the window of a pastry shop at cakes that are not for him. Sometimes I felt in myself infinite resources, the ability to succeed at everything I could undertake. Immediately, my self-esteem suffered from noticing that the course of things had taken little account of my absence. That, now, everyone went about his business as though I had not returned. I was surprised that I did not occupy a more important place in the world, that I was not offered the most attractive posts, that I was not beset by the most desirable girls.

Doubtless, one is always alone. But not to the same degree, not in the same way. In Africa, it was the sudden break with my whole past that had given me a sense of isolation. Returning to Paris, I should have found what I had been missing. But it did not work out that way. I felt out of step. I felt I did not matter a great deal to anything or to anyone. Loneliness had become a sort of natural setting, an element; and in it, I immersed myself, as both island and ghost. Perhaps I was missing the fellowship of fighting men. Perhaps I envied those who were active in a political party and could say "we." But I had an aversion to parties and their lies. For several months, the conquerors had remained isolated by their victory, the conquered by their defeat. The time came to reunite. To smooth things over. Which revolted me. One evening near the Opéra, I entered a café. It was the hour when the colors begin to falter, to free themselves little by little from the sun before being lost in the oncoming evening: as the fish that one throws back in the river take on the color of the water before disappearing in it. Through the window of the café, I looked at the crowd going by on the boulevard. I was trying to identify people. To type them. To search their faces for signs. Signs of the hangman and of his victims. Clues to the torturer and to the tortured. To those who had fought the Nazis and to those who had done business with them. But everything was leveled, equalized in the evening's grayness. Nothing but smooth and neutral faces. All these people passed one another, ready neither to flee from each other nor to come together. The world was submerging the horrible tragedy that had lasted five years, closing over it like water over a stone. Then what would joining a political party be but a coat with holes? An illusion flung over loneliness.

Twice I had tried living with a girlfriend; one a student of literature, the other an aspiring actress. The first, small; her face half hidden behind a curtain of black hair; tormented, possessive. The second, tall; tousled blond hair; smooth clear skin; without the least anxiety. The former was constantly asking, "What are you thinking about?" The latter was constantly repeating, "Guess what I'm thinking about?" Each in her own way, but both with the same enthusiasm, wanted us to plug our souls into each other, like telephone switchboards. With the first, I lasted six days. With the second, four. My lone companion at the time, the only one with whom I could go for a walk in the evening along the Seine without either one of us lying to the other, for we had taken similar paths and found ourselves in similar circumstances, was Yves C. Blond, thin, elegant Yves was the son of two well-known musicians. A medical student, he had escaped from France through Spain and joined the DB in Morocco in 1943. After the war, he, too, had completed his medical studies as soon as he could. He, too, had decided not to practice medicine. But he, at least, knew what to do: filmmaking. After shooting a film about the Second DB, he was learning his craft as an assistant director.

It was with Yves that I had, after the armistice, done some reporting about the Nazis hiding out in Austria. I had also accompanied him to Switzerland for the showing of his film about the

DB. A journey out of time in a surrealist landscape. A hallucinatory world, with houses intact, nights full of light, shop windows full of goods, trains that arrived on time in stations with glass windows. A country that had an air of unreality: where everything was in place; where the inhabitants showed no surprise at this abundance and continued the same life with the same slow, sad gestures. Back in Paris, we tramped the streets in search of new truths, in pursuit of mysteries that the city hides behind stones and creatures. We were nothing. We wanted everything, bearing within us all the dreams of the world. We avowed our intoxication with grandeur, our demand for purity and hardness, in the innumerable film scenarios we constructed at night in cafés and bars. The next day we would hurry to deposit these scenarios at the Authors' Society to protect forever our rights in these masterpieces. "The Last Fairy," for example, described an Olympus where the gods are sick of humans, disheartened by their turpitude, and decide to make a end of them once and for all and to destroy the earth. At the insistence of Aphrodite, however, they agree to send down to earth one last envoy: a fairy whose mission is to see whether there still exists in this debased humanity the slightest hope of brotherhood, the slightest trace of selflessness, of generosity. For a whole week, the fairy explores the cupidity and rottenness of this world: going from disappointment to disappointment; suffering failure and humiliation; sadly packing her bags to return to Olympus. When, suddenly, descending the stairway to leave, she runs, by the greatest of chances, into the last example on earth of human purity, the last paragon of innocence and idealism. The earth then can continue its round. Mankind is saved. We hatched at least a good dozen scenarios of the same stripe. Fortunately, none of them ever went farther than the cardboard boxes of the Authors' Society.

I had known her for scarcely a month. I had seen her three times, perhaps four, when I had to leave for Copenhagen to attend a congress of microbiologists. And yet, as I was taking off from

Paris, I knew that I would miss Lise, that I already cared for her. In the plane, I tried to reconstruct her face, her dark eyes above blooming cheeks, her tapering nose under a mass of hair. But the shapes broke up as they appeared; blurred to come back together in new designs, as if her face was a skein of mobile lines that my gaze wound and unwound in endless configurations. As, over the Baltic, the plane descended against the light, the sun reflected on the gray sea, wreathed the city of Copenhagen in gold. I wanted to feel Lise near me, to share with her the blaze that precedes the onset of night. When I arrived in Copenhagen, I felt this unknown city breathe, felt it inspire in me something like a desire for happiness. Over the next few days, I assiduously attended the congress. As soon as the sessions were over, I went out to look around the city. But everywhere, in the meeting rooms of the congress as on the avenues, I saw Lise's small head, her thick brown hair swinging with the rhythm of her step. Copenhagen became a stage set up to receive the image I projected on it. I explored this new city with her. I walked with her on the quays, by the cargo ships lying at berth. With her I strolled past the shop windows and along the paths of the Tivoli Gardens. With her I stopped at a bistro in the port to taste smoked eel and Danish beer. With amazement I discovered within myself an immense desire for shared tenderness. This girl, whom I still hardly knew, occupied all my thoughts. As if, in bringing me a theme, she had restored my taste for dreaming.

We had met at a concert, which Lise was attending with her parents, I with a friend from Free France who knew them. Her family, M. and Mme. B. and the two children, had left France in June 1940 to take refuge in the United States. When he was twenty, Lise's brother Jacques had joined the Free French. After some months of training in England, he had been posted to an infantry unit of the Second DB. On the very morning of his arrival in Alsace in January 1945, Jacques was killed with the first shell. For Lise and her parents, the return to France, immediately after the war, had begun by a pilgrimage in the footprints of their lost son and brother.

In the hubbub of a café where we had run in to each other after the concert, Lise's parents and my friend, André J., talked to each other about the war, politics, the United States. As for ourselves, Lise and I got to know each other. As always when meeting a girl, I felt awkward and dim-witted. Lise appeared rather more at ease. She asked me about Free France, my family, my work, my hopes. All this while digging into an enormous scoop of chocolate ice cream topped with a mountain of *crème Chantilly*. She was beautiful in her green dress with a black pattern. Tall, slender, graceful, with a small waist and a mass of long hair, her head on a long, supple neck straight out of the Florentine Renaissance. While polishing off the ice cream, licking a bit of the cream left on her lips, Lise was casting sidelong looks at me, apparently sizing me up. Then she began telling me about herself. Describing her arrival in America. How, at fourteen, she had suddenly found herself transplanted. Her initial refusal to speak English, to adopt new habits that had little to do with her past. Her difficulty in making friends with young American girls. The high school in New York. Then Bryn Mawr College: the cliques, the groups, the alumnae, the whole "scouting" atmosphere she had little use for. Then, a French university, a New York branch of the Ecole des Hautes Etudes in Paris, where she began a major in philosophy, with well-known teachers, professors from the Sorbonne seeking refuge in the United States. And finally music: the piano; harmony learned with a group of famous Jewish musicians from Eastern Europe who had fled the Nazis. In speaking of music and her musician friends, Lise became animated. Her face took on an inward fire, as if a night light had suddenly been switched on, giving it color and contrast. She had known in New York the two American musicians who had played Bach sonatas for violin and harpsichord at the concert that evening. Particularly the violinist, with whom she had had several opportunities to play music.

With her head bent, the dark mass of her hair stood out clearly against the leaves of a bush that adorned the café. She smiled. I was amazed to see how much her smile transformed her face. It became round. Her teeth were so gleaming that I could hardly take my eyes off them. She spoke with great warmth and energy, as though seeking to persuade me. I sensed in her a sensitivity, a tremulousness that I had not yet seen in any woman. Except, once perhaps, in Odile. I suddenly decided that, except for her long hair, Lise looked like my mother. The same way of holding her head. The same regularity of features. The same direct gaze. Whether real or imagined, I could not get this resemblance out of my mind. The association of her with the memory of my mother suddenly infused Lise's person with new sweetness and tenderness. Allowed her face entrée to a world of dreams to which she might otherwise not have had access.

In the days that followed, our dialogue continued on several outings: a walk through the Bois de Boulogne, a visit to the Jeu de Paume Museum. In America, Lise had been a great museum goer. She often sought refuge in one for an hour when life at school or at college had become oppressive. She had her favorite paintings in each museum: the Rembrandts in New York, the El Grecos in Washington, and so on. At the Jeu de Paume, Lise stopped for a while before a Cézanne: *Apples and Oranges*. She commented on it in her low voice, with a visible pleasure which she tried to communicate to me. She spoke of order and harmony in the colors, the tonality of the reds and yellows. She spoke of resonance and of fugue. In describing this painting, its coloring, its composition, she used terms from music. I could not help smiling, remembering how, a few days earlier, after the concert, Lise had done just the reverse. In describing the music of Bach, she had used terms like *texture, coloring, arabesques*. She had pointed out the richness of contrasts, the architectural precision of curves of endless variation. As we left the museum, the sun sparkled over the Tuileries. It was hot. Strollers sought the shade of the ancient trees, so close together that their leaves stretched overhead in a thick crown, a green canopy rippling in the breeze. On the ground, splashes of sunlight seemed to dance among the shadows of the cool and flickering leaves. We walked without speaking among the playing children. And our silence attested to an understanding acquired, by her as by me, since our first meeting.

Yet she also had a mind of her own, which manifested itself one evening when we had been to the movies to see *The Devil in the Flesh* with Gérard Philippe and Micheline Presle. Lise loved it. I did not. Each tried to convince the other. Our first argument. For a moment, our voices were raised. Just time to discover that Lise's look could suddenly change. That her eyes, which in repose were almond-shaped, contracted and grew round when she was irritated. A brushfire extinguished in shared laughter. But her display of firmness had pleased me. I liked Lise's whole personality, her need for the absolute, just the thing to attract someone of my sort, scarcely recovered from the war and eager for the ideal. When I told Lise of my imminent departure for Copenhagen, she gave me to understand that time would hang heavy on her hands until my return. She thus provided me, as a going-away present, with a theme for endless reveries. In a few days of separation, I was to invest her with everything the human imagination can put behind a face.

When I returned, I associated the thought of Lise with all my dreams of happiness. It was not simply the attraction of the first days. It was a true need to love that impelled me toward this face with its gaze, its smiles, the movements of its mouth. The Paris I traversed with Lise in the early summer sunlight seemed to me a new Paris, sparkling with light and with joy. Several times, going to see her, I found her at the piano. There she shone. She showed what she was made of. I learned with her more music in a few weeks than during my whole previous life. On the evening of Bastille Day, Yves C. took us in his Jeep on a tour of the neighborhood street dances. A few months later we were married.

Feeling one's way. Discovering oneself. Deciding what one wants. What one is capable of. Finding oneself a truth. Not lying to oneself. Not playing any more. Escaping boredom, comedy. Coming out in the open. Committing oneself, finally, to some activity. When one looks back, life appears, in the end, to have unfolded. But let one seek to give it a little presence, a little animation, it fades away among memories. The reason is that the endless inward discourse is lost. The voice is no longer the same. No chance to restore it. This person who was speaking, who was speaking to you, escapes. He vanishes. One may well dwell, with rage or tenderness, on the image of the vanished one. One has tried in vain to put oneself in its place, to experience its hopes and disappointments, to recover its habits and its quirks, to revive its passions and its *idées fixes*, its longings and its fears. How to prevent oneself from perceiving its behavior in the light of results it could not foresee, of facts it could not know. How to avoid giving particular importance to events whose effects, later, weighed heavily on its life but were at the time lived through casually. How to reconstruct a moment of the past without making the future become more real than that past as it was occurring? When a life advances, it is the end that endows the beginning with its truth.

The decision to plunge into research, for example, to take the step despite unpropitious circumstances, probably developed little by little, slowly maturing before it bloomed in consciousness. Nonetheless, it now seems to me to have been made in a flash, while dining with a cousin who had just committed himself to this path. For too long I could not persuade myself to leave the Penicillin Center. It was the Center that let go. It cost a lot and brought in little. It shut down. For some months I worked in a small pharmaceutical business. Charged with improving two or three medications of little value, I found the work of no interest. I quit. I then remained for a few weeks without job or activity, champing at the bit.

The only thing that tempted me was biological research. But I measured too well the gaps in my knowledge to try my hand at it. Dreading mediocrity, I shrank from taking the plunge. And then, one evening, a revelation. A dinner for four in the Latin Quarter with Lise and her cousins, the recently married Francine and Herbert Marcovich. Tall and strong, with hazel eyes and an open face, Herbert loved to laugh and to tell bizarre anecdotes. His story was similar to mine: medical studies begun in 1939; fighting the war with the army in Africa; medical studies completed in haste; then the refusal to practice medicine. But there the resemblance ended. For Herbert had dared to go from medicine to biology. He worked on the rue Pierre-Curie in a genetics laboratory directed by Boris Ephrussi, where he studied certain mutations of yeast. Amazed, I gazed at this person as if he held the truth. What until then had seemed to me impossible because of age, ignorance, and incompetence, he had calmly effected. To hear him tell it, we were not too old to begin doing research. What we lacked, we could always make up by getting science certificates at the Sorbonne. The only problem was to be nearly thirty years of age, with too much military service behind one, able to muster the courage to return to university classrooms with youngsters of eighteen or nineteen; to return to taking courses and exams. I listened agog. This guy was my own age. He knew no more than I did. He did not seem more gifted than I. And he had had the nerve to do what I wanted to do! With entire calm, without the slightest inhibition. Better, he seemed to have succeeded. Now he was working in one of the best laboratories in biology. He did experiments. He took part in the common life, in the common effort. I couldn't get over it. As Herbert spoke, I felt an excitement rising like a storm. If a man of my generation could still go into research without making himself ridiculous, then why not I? I came home all stirred up. Lise, seeing my excitement, felt that, for the first time perhaps, I might be able to overcome my despondency, and encouraged me. Overwrought, I slept poorly that night.

In the morning, it was decided. I had become a biologist, born and bred. I felt the sign that henceforth I would carry in me. I had unsheathed my sword. I was going to forge my destiny. I was headed for glory. I saw myself in a laboratory. Alone, I thought up and performed a series of experiments that at first did not look like much. Gradually, however, my results developed, arranged themselves, gained in depth. They overturned the biological landscape, modified the problem of cancer. Unknown at first, their importance eventually compelled recognition. It was a success, a triumph. People sought me out to congratulate me, to acclaim me. Lise was surrounded, fawned over. Very quickly, however, I had to hide to escape the social world, for I had better things to do, other ideas to carry out. I refused to waste time on trifles. Nevertheless, behind these bursts of euphoria, there lingered a nagging anxiety. The fear of having no talent; of being good for nothing.

Before applying myself to biology, I had to answer several questions. What kind of biology? Where? With whom? On the first point, I had a little idea. An idea formed from readings, lectures, discussion with scientists I had met here and there. At a given point in any science, there are lively fields and stagnant ones. A few years later, things are different. But, once embarked in a certain direction, it is hard to change. I had, therefore, not to make a mistake. It was a matter of intuition, of "nose," as much as of logic. There were certain indications that soon there would be some commotion in the interface between genetics, bacteriology, and chemistry. Of all the possibilities in biology, it was genetics, the science of heredity, that most appealed to me. First, because heredity lay at the very core of living beings, determining not only their shape and properties but even their very metabolism and chemistry. Also, because genetics had the virtue, rare in biology, of being a quantitative science. Finally, because genetics had just acquired new importance, even prestige, following the incredible Lysenko affair. That a charlatan could, in the middle of the twentieth century, obtain the full support of the powers-that-be in a country—government, party, press, courts, police—to impose an idiotic theory on biology and a catastrophic practice on agriculture, was indeed incredible. That an unscrupulous careerist succeeded in getting prohibited both the teaching and the practice of one of the most solidly established sciences was incredible. That one could argue against Soviet geneticists not on the basis of experiments but from the texts of Engels, and criticize genetics for its incompatibility not with facts about heredity but with the tenets of dialectical materialism, was incredible. That a political dictatorship did not hesitate to deport and imprison scientists on the pretext that they had practiced a bourgeois biology serving to support a reactionary politics was incredible. That ideological passion could lead to the self-abasement, the servility of men as free of all apparent constraint as the French communist intellectuals grouped behind the poet Louis Aragon, was incredible. In the face of this collective lunacy, genetics became a bastion of reason. To do genetics was to say no to intolerance and fanaticism.

The only biological material familiar to me were bacteria. Up to then, their tiny size and their properties seemed to exclude them from the laws of genetics. At the congress in Copenhagen, however, I had met an American researcher who had announced some news in this field. He spoke of mutation, of copulation, and even of what he called the transformation of a bacterium by a nucleic acid extracted from another bacterium. Nucleic acids meant practically nothing to me, except that Jean Brachet had also stressed their importance in his book on chemical embryology. Without understanding much of it, I scented the promise of an imminent turmoil in this field. Why, then, not try to worm my way into it? In embarking on this new life, what a strange mixture of ignorance and a good "nose"!

And also of luck. For with my inadequate knowledge of the scientific seraglio and its ways, I began clumsily. In fact, I reversed the order of things. According to the rules of the game, I learned later, the student first seeks out a supervisor. After which, if all goes well, the supervisor obtains a post for his pupil in a research organization. But in my fear of spending my life on the fringe of things, in my anxiety to find, finally, at nearly thirty years of age, a social motive and a means for earning my living, I first went after a post. My first visit: to Professor Terroine, director of life sciences at the National Center for Scientific Research (known by its French initials CNRS), whom I had met two or three times when he had been chairman of the Scientific Council at the Penicillin Center. Tall, elegant, affable, Emile-Florent Ter-

roine resembled Léon Blum, a resemblance he carefully cultivated, as much by the shape of his mustache and wide-brimmed
hat as by his rather mannered and veiled way of speaking. So I
approached him to speak of my ignorance, my willingness, my
desire to devote myself to genetics. He received me with courtesy.
Listened to me with kindness. Brushed me off with civility.

Second interview: with Professor Louis Bugnard, director of
the National Hygiene Institute, whom I had never met. I found a
tall, massive man with slow gestures and a charming smile. He
received me courteously. I revealed to him my ignorance, my
willingness, my desire to devote myself to genetics. Graciously,
he asked who my supervisor was. Unfortunately, the war years
and the African campaigns hardly lent themselves to the hunt for
a supervisor. With extreme kindness, Professor Bugnard then
explained that he could do nothing for me. And that, moreover,
genetics did not interest him.

Third interview: with Professor Tréfouël, director of the Pasteur Institute. Although I had not met him either, I knew he was
a father of the sulfa drugs. With a round pink face and beautiful
wavy white hair which made him look like an American senator
in some Hollywood film, Jacques Tréfouël received me courteously. Without even giving me time to tell him about my ignorance, my willingness, and so on, he explained to me the importance of sulfa drugs. Graciously, he asked me about my
campaigns; but before I could reply, he went on to describe the
horrors of the war at the Pasteur Institute, which the Germans
had tried to take over. In vain, he added, with a grin. Then, very
pleased with our interview, he immediately offered me a research fellowship to enter the Pasteur Institute at the beginning
of the next academic year, in October 1949. The first year, I
would have to take the "Great Course," the introductory survey
course in bacteriology, virology, and immunology. After a while,
I would have to find, among the departments of the Pasteur
Institute, a laboratory in which to exercise my talents.

I left the institute intoxicated with jubilation and pride. The
Pasteur Institute was the mecca of biology; the seat of innumera-

ble discoveries; the laboratory of eminent researchers. The dream was gathering substance. The only shadow in the picture was the small size of the research fellowship. All the more serious since, a few weeks earlier, Lise had given birth to a marvel of a little blond boy, named Pierre, whose bright blue eyes were beginning to survey the world in wonder.

On the first of October I was on hand at the Pasteur Institute to take the Great Course, a full-time combination of theoretical lectures and lab work. Still remaining was what seemed to me the main thing: to find a laboratory and a supervisor whose work was oriented in the direction I was hoping to take. I got some ideas concerning the landscape of French biological laboratories from two friends, Jean Zieglé and Norbert Grelet, who had worked for a while at the Penicillin Center. Through them, I learned that, in my field of interest, there were in Paris two laboratories of exceptional caliber: one was headed by Boris Ephrussi at the Institute of Physicochemical Biology on the rue Pierre-Curie; the other, in which Jacques Monod was working, was headed by André Lwoff at the Pasteur Institute. As a neophyte Pastorian, I was of course eyeing the latter. I had met André Lwoff at the Penicillin Center and Jacques Monod at a meeting about Lysenko. The younger and shorter, Monod appeared less intimidating. I went to see him. But, at my request to work with him, he sent me to see Lwoff. Therefore I made an appointment with Lwoff. On the agreed day, I arrived early as usual. The laboratory was on the third floor of the chemistry building, under the eaves. To reach Lwoff's office, one had to traverse a long meandering corridor bristling with apparatus of all kinds. When I arrived, Lwoff was lunching in his laboratory in the company of two charming women, his technician and his secretary, I later learned. In his office, I told him about my ignorance, my willingness, my desires. He fixed me for some time with his blue eyes. Tossed his head several times. His laboratory was already fully staffed. There was no place for me.

What a disappointment! But I was stubborn. In the course of that winter, I came back several times to see André Lwoff. And received each time the same gentle blue-eyed gaze, the same tossing of his head, the same refusal. In the spring, my optimism began to fade. I went in search of an alternate laboratory just in case. In June, I risk one final attempt on André Lwoff. Arriving, I find his eyes bluer than usual, the toss of his head more pronounced, the welcome warmer. Without even leaving me time to express, once again, my ignorance, my willingness, my desires, he announces, "You know, we have just found the induction of the prophage." I greet this news with an "Oh" into which I put all possible surprise, amazement, and admiration, while thinking, "What does that mean? What can a *prophage* be? What does it mean to induce a prophage? What language is he speaking?" He looks at me again for a while, his chin touching his chest, and asks: "Would you be interested on working on the phage?" Staggered, I can only stammer, "That's just what I'd like to do." "Then, go off on vacation and come back the first of September."

I go slowly down the stairs, delighted and disoriented at once, like a child who has finally received a long-coveted toy. At the same time, this story about the prophage worries me. I dash into the first bookstore I see to consult a dictionary. *Induction? Prophage?* When I leave, I am none the wiser. And then, I cannot stop wondering why André Lwoff has admitted me into his laboratory. Perhaps owing to my obstinacy, or to my less-than-classical training. Or perhaps, more simply, thanks to his good mood as a result of the "induction of the prophage." To this day, I still do not know the answer. What I do know, however, is that, had I been he, I would surely not have accepted into my laboratory a fellow like myself.

VI

GAINING ADMITTANCE to this laboratory was a mere start. I still had to get myself accepted by the group, to become a member of the clan. I had to penetrate this unfamiliar world, to discover its rules and its hierarchies, its folklore and its language. I had to organize this world around myself, to enter it and find my place. For a time, I had a sense of remaining at the door, observing the spectacle from outside, as if I were merely a visitor. I understood little of what the others were doing, little even of what they were saying. I should have liked to play with the new words, to relish the technical terms, repeating them over and over, combining them ad infinitum, as I had in my childhood. But apart from the term *enzyme*, whose closeness to *azyme* and the unleavened matzoh of the same name enchanted me, the words in the laboratory eluded me. I found in them nothing to grab hold of, nothing to hook on to. The words whirled around me in a sort of frenzy, but remained colorless, devoid of meaning. They seemed to be mocking me, defying me. As though they were guarding the portal to a temple to which I would be refused admittance.

Shortly after my arrival, I attended my first seminar. The seminar is to research what the blackboard quiz is to the lycée student: an obligation to give a demonstration in public, before one's peers, of one's ability or its lack. For the research scientist, it is not enough to construct a conception of reality in bursts of theories and experiments. One also has to make it known, to introduce it, by force if need be, into the minds of one's colleagues. There are many means to this end: journal articles, congresses, colloquia; above all, seminars, true rites of initiation, free-for-alls conducted before specialists little inclined to adopt a competitor's ideas or results. That day, an American biochemist, Sol Spiegelman, was presenting his data on "enzymatic adaptation" in yeast. Enzymatic adaptation was much talked about in the laboratory: it was the subject of the research of Jacques Monod and his team. I no longer associated the expression solely with addiction to matzoh, for I had read a recent article of Monod's. With some bacteria, certain enzymatic activities could be detected whatever the condition of a culture. Other activities, in contrast, varied with these conditions. With the colon bacillus, a common intestinal bacterium, for example, one always found the set of enzymes needed for the consumption of glucose, even when, owing to lack of glucose in the medium, these enzymes were totally useless. On the other hand, the enzyme used by the same bacterium to consume another sugar, lactose, appeared only when needed, in the presence of lactose. The enzymes for glucose were called "constitutive." The enzyme for lactose, beta-galactosidase, was said to be "inductible" by lactose. This power of the bacterial cell to modify its chemical activities as a function of the environment testified to the flexibility of living beings, to the cell's capacity for adjusting to varied conditions. The mechanism for this capacity was unknown. The problem was to discover it.

The seminar was held in André Lwoff's office. A small attic room with a sloping ceiling, walls lined with cupboards and bookshelves. In one corner was a movable blackboard. From a desk in the center, an inscrutable André Lwoff presided over the session. The scientists of the lab and some visitors were seated around him on chairs or perched on stools. At the time, there were only a few experts on enzymatic adaptation, perhaps eight to ten researchers throughout the world. And most of them were attending this seminar. Besides Monod, there was the Englishman Martin Pollock who always, when speaking to you, kept his handsome, insolently aristocratic face fixed on the horizon above your head; the Canadian Roger Stanier, a debonair giant with a wide grin, his eyes crinkling with mischief behind his glasses; Melvin Cohn, an uninhibited young American who had come to work with Monod for a few years. In short, along with the speaker, the cream of enzymatic adaptation had gathered that day in this room.

For the day's speaker also was a champion of this specialty. Short, stocky, broken-nosed, looking like a featherweight boxer, Sol Spiegelman spoke as much with his hands as with his voice. His gestures abrupt and precise. His sentences short and choppy. But above all, alas, he was speaking English in a nasal American drawl that I could not fathom. My English was nothing brilliant. Laboriously acquired in the lycée, it had improved a bit during the war, in Great Britain and with the British Eighth Army in Africa. As for the Americans of Patton's army, I had had too little contact with them to become familiar with their accents. As Spiegelman talked, I discovered, to my grievous disappointment, that I could not follow what he was saying. One more failing! This one unexpected! For some minutes, I tried to hold on, to grasp some scraps here and there. I tried to get a foothold. I concentrated. In vain. Despite my best effort, I was able only to catch a word here, to snatch an expression there. Vague glimmers in the night. Too infrequent traces of sense in a series of growls, in a torrent of sounds without meaning. A buoy sometimes, when Spiegelman, at the blackboard, wrote a formula, drew a diagram. Very quickly, however, I gave up. I tuned out, content to follow the spectacle from the outside. To observe the actors. To examine the audience. To decipher the ritual.

For Spiegelman's talk did not unfold without incident or sudden stops, like the majestically flowing rivers of oratory in the lectures at the medical school. The audience interrupted incessantly. Breaking in with more or less well-meaning remarks. Chopping up his discourse with questions to which Spiegelman briefly and confidently responded, before retrieving its thread. Sitting in the first row, the specialists of enzymatic adaptation would not let up, continually badgering him, provoking him, nipping at his heels like excited puppies. They talked among themselves in an undertone; they laughed; they made signs to one another. The most aggressive was Melvin Cohn, always ready to quibble about some experimental detail. The sliest was Pollock, who asked irrelevant questions to give the impression of not understanding. The most imperial was Monod, who in impeccable English endeavored to show Spiegelman that he was not asking the right questions, that he should have done other experiments. Saying not a word, the always imperturbable Lwoff never stopped playing with his ruler. As for Spiegelman, he was not easily confounded. Thick-set, his head sunk into his shoulders, lips pursed, he listened to comments and criticisms while playing with a piece of chalk in the fashion of a Hollywood gangster with a coin. Then he gathered himself together and, with renewed spirit, returned to the attack, describing a new experiment which he summarized in a diagram on the blackboard. An experiment immediately subjected to a barrage, a running fire of questions and criticisms from all directions. Suddenly I experienced a feeling I had had a few days earlier. During our vacation, Lise and I had attended our first bullfight, in the arena at Bayonne. Fine little bull this Spiegelman, very excitable, very combative. Always ready to charge toward the red cloth that the elegant toreador Monod was flourishing so audaciously. From time to time, his assistants Pollock and Cohn sprang forward to implant their banderilleros. And at each new pass, the aficionados of enzymatic adaptation cried *Olé*!

Neither severed tail nor ears were displayed. The deathblow came later in a café on the boulevard Pasteur, among glasses of beer. Toward the end of his talk, Spiegelman had presented a theory of which I understood nothing. What I did understand, however, was that in the café the theory was dissected, torn apart, shredded into tiny pieces. Bit by bit, the bull weakened. A final thrust of Monod's *descabello*. The bull's final spasm. And resistance ceased. All this amid laughter and joking. Jokes in English, unfortunately, with a variety of accents. The exchanges went over my head. What irritated me most was my inability to spot a change of subject, to realize that where they had been talking about a Hitchcock movie they were now discussing Marxism, or a Bach cello sonata, or the cost of sailboats. Sallies spurted from every direction. With, suddenly, a brief exchange of fire between Pollock and Monod. Impossible for me to follow the details. I just grasped that they were discussing the theater. English drama versus French drama. Shakespeare versus Molière or Racine. Two trained roosters beak to beak, vying with humor and sarcasm. Not because of nationalism, but through a will to power, through a desire for intellectual domination.

It was a surprise, that first seminar. I left both dazed and fascinated. This universe of research, still charged with mystery, suddenly appeared in a new light. This was not the cold, studious, stiff, slightly sad, slightly boring world one often imagines. But, on the contrary, a world full of gaiety, of the unexpected, of curiosity, of imagination. A life animated as much by passion as by logic. With its coteries and its own community, its idols and its taboos, its codes and its rites. With its struggles, too. The competition for primacy: not for temporal power but for intellectual power, dominance over the group. On the stage of enzymatic adaptation, the dominant figure was unquestionably Jacques Monod.

To do or to learn? To launch, head lowered, into research without bothering with university training? Or, on the other hand, to acquire first at the university the basic elements so as to be able, later, to go on to the real work? For a normal student, this problem does not arise. In university classrooms, he familiarizes himself as a matter of course with the fundamentals of biology before entering a laboratory. But my scientific culture was practically zero, being limited to the residue of my far-off medical studies as well as some scraps gleaned from journal articles and lectures during my stay at the Penicillin Center. I had no overview, no idea of the big problems or the hypotheses then in vogue. Solid theoretical training thus did not seem a luxury. On the other hand, I was already thirty. I had as yet done nothing. To lose another two or three years seemed madness, especially as I had not come into this laboratory to chase butterflies or to study for a college diploma.

Fortunately, the question did not preoccupy me for long. The research being done in the laboratory was so exciting, the atmosphere so attractive, the enthusiasm of my boss, André Lwoff, so infectious that the question soon receded from my mind. In these always difficult beginnings in a new profession, in a new world, it was Lwoff who gave me confidence in myself, who helped me to vanquish my doubts about my aptitude for research. One day, in spite of my shyness toward him, I resolved to ask him about the necessity, or even the utility, of taking the time to pursue certificates at the Sorbonne. He gazed at me with a sly smile: "If you want to have a university career, then this is surely not the place to come to. Go to the Ecole Normale or take the *agrégation* instead." Then, after a pause: "Don't lose sleep over it. Do experiments, and the rest will come on its own." Then, after another pause: "If you insist, go after one or two certificates, biochemistry or genetics. But on the side, extra. Don't let it keep you from working." And so I did. Some weeks later, when I was beginning to get results, I hurried to show them to him. "So!" he said with a grin, "you see? Research is mainly a matter of flair. But also of tenacity, stubbornness. Faults that, judging from your war record, I gather you are not totally deficient in." And since then, I have tended to class certain research procedures with military operations: the will to conquer; applying a strategy and a tactic; the necessity of choosing a terrain; developing a plan of attack; concentrating one's forces on a particular sector; focusing and modifying the initial plan according to the reactions obtained. In short, going on the offensive on all fronts.

At fifty years of age, André Lwoff was already at the pinnacle of his career. While he enjoyed an enormous reputation abroad, he was unknown in France, the reason being that he had not received the official sacraments. The son of a psychiatrist, he had begun to study medicine. Attracted early on to research, he had, at twenty, entered the Pasteur Institute in the laboratory of Félix Mesnil, a former student of Pasteur. And for a time, he had collaborated during the summer months in marine laboratories in Brittany and the south of France with a great protozoologist, Edouard Chatton. In the 1930s, Lwoff had done a stint in several foreign laboratories of the first order: in Heidelberg with Otto Meyerhof, then in Cambridge with David Keilin (two of the best biochemists of the time). Lwoff had few students. He loved to work with his hands, surrounded by his wife, Marguerite, by a technician, and by one or two collaborators. He detested large teams, industrial science where in the morning the boss gives instructions to an army of assistants and comes back that evening to pick up the results as a tycoon stops by to pick up his profits. For Lwoff science was sensual. It consisted, first, of objects, of organisms or fragments of organisms which had to be manipulated, fiddled with, felt. Later, but just as important, came the words, which likewise had to be caressed, rearranged, so as to describe things with finicky care.

Tall, thin, very upright, André Lwoff dressed simply but elegantly. He was a man of refinement in all things: in his dress as in his reading; in his menus as in his distractions; in his wines as in his language or writing. He loved short sentences and did not hesitate, in conversation, to correct what he considered mistakes in French. He ran a fine comb over each publication coming out of his laboratory, to ferret out the improper word, the misplaced comma, laboratory jargon, and especially *franglais*, then in full swing. Carried away by enthusiasm, he ended by caring little about content as long as he could control its form.

Like the great lords of old, André Lwoff had his friends, his people, his "clientele," who were always right and could ask anything of him. Lwoff never stinted them either his help or his friendship. He had confidence in them to the point of credulity. From the moment I crossed the threshold and was admitted into the clan of his pupils and friends, he showed me an unfailingly affectionate benevolence. He never ceased to encourage me, to show off and to praise my work, to speed up promotions. And when I had grown up, when I began to make a stir, he continued, without taking umbrage, to push me forward. A rare virtue in a world motivated by ambition, by the need to shine, to be recognized. He looked at me, I believe, rather like a son, while I looked on him rather like a father.

Around André Lwoff, an exceptional ambiance prevailed, a warmth and openness of mind, of life and imagination. Someone who has not lived and worked in the attic might not appreciate the richness and generosity of a personality who often retreated behind a somewhat haughty reserve. But Lwoff was all of a piece. He never bent to either compromises or half measures. Intrigue repelled him as much as flattery. What he had to say, he said plainly and without circumspection. So he had enemies: all just as staunch, as loyal as his friends and whom he treated with an irony, a hostility in proportion to his kindliness to the latter. In some, he excused everything; in others, nothing. With people he considered incompetents or charlatans, especially if they occupied a high civil service or university post, he could show a ferocity without concession, a candor without restraint, but often mingled with humor, even sarcasm. As, for example, in a polemical argument against a zoologist about certain aspects of evolution. In Mesnil's laboratory, Lwoff had studied the nutritional needs of various protozoa. When an organism finds in its food some compound necessary for its growth, it often benefits in ceasing to produce the compound and sometimes even loses the means of performing this synthesis. Lwoff then had the idea of relating the morphological evolution of protists to their capacities for synthesis. He had thus noticed that the alimentary requirements of protists could be thought of as an aspect of a

biochemical evolution leading the most independent forms—
those with the most highly developed capabilities for synthesis
—toward the most dependent, which behaved like parasites. In
other words, this evolution corresponded to the gradual disap-
pearance of biosynthetic processes; and in his doctoral thesis
Lwoff had named it "evolution through loss of functions." A
hypothesis that was not to everyone's liking; that had even been
severely criticized by a French evolutionist, one who was, however, universally acknowledged to be incompetent. This man saw
evolution as always proceeding in an upward direction; as pro-
gressing linearly; as always resulting in a gain, never a loss, of
function. To which, in his *Physiological Evolution*, Lwoff responded
by wondering how "this distinguished colleague would find a
way to accomplish his important work if, to survive, he had to
remain all day exposed to the sun in a solution of potassium
nitrate."

But his humor could sometimes become caustic, his candor
turning brutal and cutting to the quick. As happened at a study
section of the CNRS, to which both Lwoff and a certain biology
professor at the Sorbonne belonged. This latter enjoyed, among
the researchers at the Pasteur Institute, a reputation as an incompetent pedant. A bastard, too. He had once been publicly
shown up by Monod as ignorant and stupid; the experiments he
had published were absurd. After which, he hastened to fire his
research assistant, who had merely been following his instructions. One day, this professor made an application to the study
section of the CNRS for research funds. According to custom, the
professor left the room to allow his colleagues to discuss his case
freely. Without shillyshallying, Lwoff gave a blunt opinion about
the value of the research project. Immediately on returning to
the room, the professor was informed of these comments. After
the meeting was over, he approached Lwoff and said, "Monsieur,
I ought to slap you!" "Go right ahead," responded the imperturbable Lwoff. "My seconds will call on you tomorrow morning."
Bewildered, the man opened his mouth, then closed it and left.

Who would have supposed the study sections of the CNRS so be so turbulent, even dangerous?

What I admired most about André Lwoff were his acute sense of the living world and the virtuosity of his approach to biology, of his choice of research topics. Governed by intuition and flair rather than by method, he practiced science like an artist. At heart, he *was* an artist. Several times he had decided to change subject, to take on new research. Every time, he had gone at the question from a new angle. Every time, he had gone directly to the heart of a key problem in biology. Every time, he had stamped a particular area with new results, new ideas. His reputation came mainly from two discoveries of the highest importance. The first concerned the status and the role of vitamins. To multiply, certain bacteria require, in the culture medium, the presence of a particular compound which other bacteria, however, can perfectly well do without. These "growth factors" necessary for bacteria often proved to be identical to one or another of the vitamins required for the growth and health of mammals, notably of man. Before the war and at the same time as the Englishman Paul Fildes, André Lwoff had shown that these growth factors, these vitamins required by certain organisms, are in fact constituents of *all* living things; that they are indispensable to *all* life; that they play definite roles in specific chemical processes in the metabolism of *every* cell. If certain organisms require the presence of these factors in their food while others can do without them, the reason is simply that the latter manufacture these compounds themselves while the former are incapable of doing so. Hence, the necessity for the former to obtain these indispensable substances ready-made.

This discovery had enormous implications. Supported by the findings of biochemical analysis, it showed for the first time that, at the molecular level, the same structures and functions are found in all living beings; that the whole living world is constructed out of the same materials. Thus there emerged a new aspect, a hidden face of the evolution of living beings, whose study up to then had been based on the analysis of plainly visible forms and features. Some years later, the analysis of the nutrition of micro-organisms, based on Lwoff's ideas, was to be taken over by geneticists, notably the American George Beadle. Thus was born biochemical genetics. His method consisted of inducing mutations with "losses of biosynthetic functions" and of studying in parallel the biochemical deficiencies and the genetic changes. A prelude to the advent of molecular biology.

André Lwoff's second major discovery was the famous "induction of the prophage" which recently had aroused my curiosity. Prophage, necrophage, anthropophage. A *phage* was, in lab jargon, short for *bacteriophage*, the eater of bacteria: that is, viruses that enter bacteria and destroy them by multiplying there. A particle of bacteriophage, or one virus, thus forces its host to form, in twenty minutes, a hundred particles identical to it. The bacterium then explodes. This *lysis* frees the hundred particles, which go off to infect other bacteria, and so on. Like all viruses, these bacteriophages are too tiny to be seen with an optical microscope. For a long time, no one could perceive them. They were creatures of reason, hypothetical constructs, whose gambols could be followed, which could be counted only by indirect methods. Only the recent development of the electron microscope finally allowed their visualization. Showed that these ghosts had not simply been dreamed up; that they had characteristic shapes; that they behaved very much as had been thought from their effect on bacteria.

One point emerged from all the literature and the seminars: seeing the particles merely confirmed what we already knew, and added little to the facts. What had made possible analysis of bacteriophage multiplication, and understanding of its different stages, was above all the play of hypotheses and experiments, constructs of the imagination and inferences that could be drawn from them. Starting with a certain conception of the system, one designed an experiment to test one or another aspect of this conception. Depending on the results, one modified the conception to design another experiment. And so on and so forth. That is how research in biology worked. Contrary to what I once thought, scientific progress did not consist simply in observing, in accumulating experimental facts and drawing up a theory from them. It began with the invention of a possible world, or a fragment thereof, which was then compared by experimentation with the real world. And it was this constant dialogue between imagination and experiment that allowed one to form an increasingly fine-grained conception of what is called reality. To this game the bacteriophage was particularly well suited, lending itself to all sorts of experiments: simple, rapid, precise. It became a model for studying not only the multiplication of other viruses but also the very mechanism of reproduction: 1, 2, 4, 8, . . . 100. Seven generations in twenty minutes! A record! Material to gratify the most exacting geneticist. These investigations were conducted by a small number of Americans. They formed the "Phage Group," also called the "Phage Church," whose pope was the German-born physicist Max Delbrück.

André Lwoff had approached the study of the bacteriophage from a different angle. With certain phages, the bacteria did not always die from the viral infection. They could survive and multiply without producing viruses. Nevertheless, in the cultures of their descendants there were always some particles of the virus: as if some sort of equilibrium had been established between bacterium and virus. This state was called "lysogeny." The lysogenic bacteria had already been studied by Jules Bordet and McFarlane Burnet and, at the Pasteur Institute, by Eugène and Elisabeth Wollman, the parents of Elie Wollman, a student of André Lwoff. Toward the end of the war, the Wollman parents had been deported and had died in the death camps. When the war was over, André Lwoff had returned to this problem with, characteristically, a highly original approach. In a few weeks, he had entirely transformed it. What he wanted to study were not huge bacterial populations, such as had always been worked on before, but single organisms. To this end, he had fished with a micromanipulator for one bacterium at a time and then tracked its behavior with the microscope. This brought to light a surprising phenomenon: these lysogenic bacteria whose ancestors once had encountered a virus retained the memory of this event. After hundreds, thousands of generations, every descendant retained the ability to produce the same virus that had infected the ancestor. In each individual, then, the virus persisted in a hidden form, in an inactive state Lwoff called "prophage." It was thus necessary to try to get the virus to leave its hiding place, to reactivate it. Which Lwoff achieved in a few weeks with two young researchers from abroad who were getting some training in his laboratory: the Canadian Louis Siminovitch and the Dane Niels Kjeldgaard. The equilibrium between bacteria and prophage, which had persisted for thousands of generations, was upset as soon as these lysogenic bacteria were exposed to ultraviolet light. After a culture was irradiated, all the bacteria lysed—that is, dissolved themselves, each one releasing a hundred particles of bacteriophage. An observation whose importance it would be hard to exaggerate: it is enough to consider the advances it made possible. It transformed ways of thinking about viruses, their origin, their relation to the cell. The production of bacteriophages after irradiation was the famous "induction of the prophage": the topic that Lwoff had assigned me to study.

It was my good luck to arrive at the right time at the right place. At the right time, for in those postwar years was born a new discipline which, using micro-organisms as material, found a way to analyze the mechanisms of the basic unit of life, the cell. The study of fungi and yeast had shown that heredity governs not only secondary characteristics like the shape of a fly's wings and eye color, but all the properties of an organism, down to the tiniest detail of its chemistry. Once metabolic reactions had been transformed into objects of study for genetics, it became possible to analyze the heredity of organisms as simple as bacteria. The very properties that until then had seemed to exclude bacteria from genetic analysis suddenly made them a particularly favorable material for the study of variation. Then it was discovered that bacteria contain genes, and even that they exhibit phenomena of conjugation which are reminiscent of the sexuality of higher organisms. And, more amazing still, that by absorbing genes from their sister bacteria, certain bacteria acquired corresponding properties, which opened an unforeseen route to the chemistry of heredity. And, to bring oneself up to date, to approach the world literature on the question, it was enough to read four or five journal articles. Notably on the statistical analysis of bacterial variation by Salvador Luria and Max Delbrück; on the conjugation of bacteria by Joshua Lederberg and Edward Tatum; on the transformation of pneumococcus by Oswald Avery and his collaborators. Similarly, the literature on the bacteriophage was limited to a dozen articles, all written by the Phage Group. A rare case where ignorance is not a vice. Where, not curbing the imagination, it can even be a virtue.

I had also arrived at the right place. This new discipline was emerging from just a few centers, most of them in the United States. In France, Lwoff's attic seemed one of the few laboratories able to take part in this adventure, thanks to its orientation and level of competence. Thanks above all to the combination of the two Pasteur teams and the originality of the men directing them. Along with André Lwoff, Jacques Monod was the dominant figure in the group. Short, with black hair, a handsome face, a determined chin, and high cheekbones, he looked like a cross between a Roman emperor and a Hollywood movie star. Ten years younger than Lwoff, he had studied zoology at the Sorbonne. Shortly before the war, at the suggestion of Boris Ephrussi, he had gotten some training in Pasadena at the California Institute of Technology, in the laboratory of T. H. Morgan, one of the fathers of classical genetics. There, he had engaged in as much music as biology. He was equally ardent in both, playing the cello as delightedly as he gazed through the microscope, organizing choral groups as well as seminars. He loved to say that he had long wavered between whether to be a biologist or a conductor. Boris Ephrussi told how, on their return from Pasadena, he had received a visit from Monod's elder brother, who wanted to know whether Jacques had a future in science as brilliant as in music. In the face of the enthusiasm of Ephrussi, who found Monod exceptionally gifted, the brother had murmured, "Good! So Jacques will not be Bach. He will be Pasteur!"

Because he was interested in growth, Monod had chosen to study it in bacteria: a surprising decision at a time when bacteria were considered of interest only in their role as pathogenic agents. Monod cultivated a harmless organism, a colon bacillus, in the Sorbonne's zoology laboratory, among an array of stuffed animals. He varied the environment, the temperature, the sugars serving as an energy source; then compared growth under these various conditions. From time to time, in this rather tedious work, an oddity would crop up. For example, an asymmetry in the organism's capacity to consume certain sugars. When the colon bacillus cultivated in lactose was transferred to glucose, its growth continued as if nothing had changed. On the other hand, when cultivated first in glucose and then transferred to lactose, the same organisms stopped growing and resumed only after a period of latency. When consulted about the phenomenon, Lwoff declared it "a typical example of enzymatic adaptation." Up to that point, Monod had been unacquainted with the expression. Like Monsieur Jourdain in Molière's *Le Bourgeois Gentilhomme* who learns he has been speaking prose, Monod thus discovered that without knowing it, he had been producing enzymatic adaptation. He never, henceforth, left it. He changed its name; he constantly renewed the subject; transformed it; made it into one of the sharpest and most precise tools for determining cellular processes; but always while digging away at the same furrow, always without a break, always deeper.

Monod made bacterial growth the subject of a doctoral thesis which he defended during the war. On this occasion, one of the professors at the Sorbonne, who headed the jury, made a historic pronouncement: "This work is of no interest to the Sorbonne." It did, however, interest Lwoff, who invited Monod to work at the Pasteur Institute. It was during the dark hours of the war. Monod went into hiding; Communist Party; guerrillas and partisans. At the Liberation, he resurfaced on the general staff of General Jean-Marie de Lattre de Tassigny. Once the war was over, he went to work in Lwoff's laboratory at the Pasteur Institute. From then on, he devoted himself utterly to enzymatic adaptation, which he renamed "enzymatic induction." A personality of extraordinary range, whose many facets included a marvelous intellectual mechanism; an interest in every field; great culture in the arts as well as in science; an absolute rigor in criticism, for others as for himself; a personal charm which he modified according to the situation, to the sex and social status of his conversational partner; the ability to concentrate on any subject and to change in a minute; a certain desire to please, even to push himself forward; total accessibility to his friends or his students; a self-confidence so obvious that it had to be hiding a basic insecurity; an immense gaiety with a laugh so infectious, so resounding that one could immediately identify it from a distance; a desire to dominate intellectually, even to terrorize, that led him to sit in the first row at seminars to bombard the speaker with questions and explain to them either what they had done or what they should have done; an enthusiasm in support of any cause he found just. In short, an ardent personality. On my arrival, I was prompted by my medical training to call my superiors *Monsieur*. Which was fine with Lwoff, but not Monod. "Please don't call me Monsieur," he said brusquely. "I am not a medical supervisor." Nonplussed, I stopped calling him anything. Which irritated him. "Call me what you like, Monod or Jacques or old fart. But call me something!" I chose Jacques.

It was hot under the eaves when I joined the lab in those first days of September 1950. The chemists organized well-studied drafts of fresh air which made the bacteriologists yell, fearing contamination. The attic stretched along a hallway onto which, here and there, the laboratories opened. Lwoff and his group worked at one end of the hall; Monod and his group, at the other end. I was installed at Lwoff's end: in a large room with a laboratory bench along one wall, cupboards, a large table in the center. I shared this laboratory with an American couple, Francis and Betty Ryan. They wanted to speak French; I, to speak English. We ended up compromising: French in the morning, English in the afternoon.

This large laboratory, with its sloping ceiling like every room in the attic, had a certain antique charm. It also had a grave defect. Its size and the presence of the large table made it the only place roomy enough to hold both teams for lunch. From one o'clock on, all the researchers of the attic would arrive, armed with a sandwich, a slice of meat, or a lunch box, to arrange themselves, one after the other, haphazardly on the stools around the table.

There were few permanent researchers and few students in this laboratory, which the Sorbonne ignored. In Monod's lab were Germaine Cohen-Bazire and Anna-Maria Torriani, an Italian woman living in France. In Lwoff's lab were Marguerite Lwoff, the boss's wife and collaborator; Pierre Schaeffer; and Elie Wollman. Elie was returning from a two-year stay at Delbrück's lab at Caltech, the mecca of the Phage Group. There, he had not only learned to work with bacteriophages but had gotten to know other researchers working in the field. For two years, he had tasted life in an American university, less hierarchical, less formal, less rigid than a European university. Above all, he had learned about a certain style of research, as exemplified by Max Delbrück. A style that injected into biology a rigor of sustained reasoning which up to then had been the prerogative of physics; which aimed especially at forming a coherent representation of the system studied; which combined ferocity in criticism with whimsy in personal relations; which intimately blended science and everyday life. With a fine, highly critical, even caustic intellect, Wollman had a passion for research and a total devotion to the Pasteur Institute. It was thanks to him that I formed an idea of the bacteriophage scene, of the methods employed, of the literature. We worked in neighboring laboratories. We talked at great length. We soon became friends.

If the attic did not interest the French university, it enjoyed a solid reputation abroad, notably in the United States. Which was important for, with the war, science had switched its center of gravity from Europe to the United States. Each year, from just about everywhere, the laboratory received professors on sabbatical and students coming to do a spell of postdoctoral work. Among the latter were such members of the Phage Group as Seymour Benzer and Gunther Stent. Another of them, Melvin Cohn, stayed on for more than five years working with Monod. Between the attic and certain foreign laboratories, there was an incessant exchange in both directions. A constant flux of letters, visiting students and professors. A ferment in the air and in ideas required for a laboratory to avoid accumulating dust and rust. In addition, our American visitors' freedom of manner and expression, their critical spirit that spared neither institutions nor hierarchy, gave the attic a particular style, quite remote from the one prevailing in most French laboratories. When Seymour Benzer asked for a key to the Pasteur Institute library so he could work there at night, the librarian almost had an epileptic fit. To be able to read and write in peace, Cohn did not hesitate to come in of an evening and set up shop in the most sacred place in the institute: the crypt where Pasteur himself is buried. All this coming and going was made possible through the help of American foundations. It was they who granted the funds to support Lwoff and Monod's research. It is thanks in part to these foundations, to their generosity, to a sort of Marshall Plan for European laboratories, that after the war many of these laboratories were quickly able to regain their activity and vigor.

In the attic, Lwoff and Monod had been able to produce an atmosphere in which mingled enthusiasm, intellectual clear-sightedness, nonconformity, and friendship. Lunch was an exceptional moment in the day, one of the rare moments when work was combined with relaxation, personal life with scientific life. First, discussions begun in the corridor were continued: about a result obtained that very morning; about a seminar; about the request for a strain by an American who wanted to do the same experiment we had done, a request that sparked fury and irony. But, very quickly, the conversation would swerve; in fact, it was constantly changing. Ricocheting off an idea, a word, to land on a journal article, a book, a film, a travel story. On politics, above all, which presented many themes to scientists, with their taste for irony and polemic. We palavered incessantly on the role and responsibility of scientists in the use of the atomic bomb in Japan; on what they had done and what they should have done; on the Cold War and humanity's chances of surviving; on McCarthyism and the witch hunts, which did not spare the scientists, for many, and not the least important, had a bone to pick with the investigating committees; about the Lysenko affair and the behavior of French intellectuals and scientists; about the light thus shed on the methods and goals of Stalinism. We also unceasingly discussed French politics, its variations, its subtleties, which our Anglo-Saxon friends found a bit too Byzantine to follow. An astonishing school of dialectics, these lunches. Everything said by one of us was immediately snapped up by someone else, kneaded, turned over, ground, passed through a sieve, and finely pulverized before vanishing in pleasantry.

And then there were the laboratory's *bêtes noires*. The whippingboys. In the most diverse fields. Among the politicians, Pierre Mendès-France was almost the only one to escape invectives. Among the university types, the woman professor at the medical school who had, without knowing a word of genetics, joined Aragon in his campaign in support of Lysenko. Among our colleagues at the Pasteur Institute, the one who loved to relate his exploits in the Resistance: "I laid out flat the first German who showed up in my laboratory!" he said. "And what happened to the second one?" he was asked. "The second? He knew my work!" Among the foreign visitors, the famous American biochemist Professor T. who, upon being taken to the Louvre on his first visit to Paris, came to a halt in front of the Venus de Milo. After walking around it at length, he finally murmured, "What an extraordinary resemblance to Mrs. T." At each repetition of this story, an enormous laugh would shake the crowd, a collective laughter in which rang loudest and deepest the laugh of Jacques Monod.

But, however stimulating lunch was, however exciting the discussions, I soon was fidgeting, impatiently, waiting for people to leave and release the laboratory. In principle, everyone had to be finished and the room vacated by two o'clock. But lunch ended with the coffee ritual, presided over by Marguerite Lwoff and Sarah Rapkine, the widow of a scientist of the Pasteur Institute, a friend of Lwoff and Monod. All the researchers took coffee. They wanted it very strong. They liked to sip it slowly, in little gulps, keeping on with their discussions and jokes. But, well before two o'clock, I had only one idea in my head: to see everyone clear out so I could finally carry out the day's experiment.

A mass of blond curls, he is seated in a little wicker chair. On the table in front of him is the lunch that Lise has fixed for him: a sliced banana, ham chopped fine, little cubes of Gruyère, a baby bottle. Pierre plunges his hand now in one spot, now in another. He swallows a bit of Gruyère. He conscientiously dabs banana all over his face. Smears his head and hair with ham. But what interests him is the spoon. Not to eat. To grasp its reaction when one speaks to it. Or when one moves it. He pushes it gently. In little nudges. As if to see what the spoon will do. Whether it will respond. Or remain still. Another little push, and the spoon falls off the table. Pierre cannot catch it. He grunts. Nothing happens. He cries out. I come over and pick up the spoon. Pierre smiles. With a big shove, he makes it fall again. I pick it up. He throws it again and bursts out laughing. Pierre and the world around him. He looks. He pushes. He hits. He cries out. He tries. He makes a great effort to see how things work. Who makes them move. He gives a blow. He repeats. He foresees. He starts over. He verifies. He tries everything that comes into his head. Constantly, he makes an attempt at doing what seems possible to him. He is seeking to understand the world and the forces that animate it.

To verify that the induction of the prophage is not limited to the one bacillus that he had studied, André Lwoff had given me the task of analyzing lysogeny in another organism. He suggested the pyocyanic bacillus. I obtained some thirty strains from the Pasteur Institute's microbe bank. To begin with, it was necessary to compare all combinations of these strains two by two to determine which of them produced a bacteriophage that acted on another. This exercise reminded me of the naval battles I had played in my youth. Of the thirty strains, a good dozen proved to be lysogenic. Some of them even became familiar companions. I recognized them from afar: the bacteria, with their blue-green or brown or pale pink colonies; the bacteriophages, with their lytic plaques of more or less large size, of more or less clouded aspect. I called them by the number I had assigned them. I had my favorites, my pets, the ones I was most likely to choose for my experiments. Every evening, I prepared them for the next morning. Every morning, I reinoculated them into a fresh medium to have them at just the right degree of growth after lunch. I spent a good part of my days this way with my strains. Cultivating them; leaving drops of one culture on agar plates seeded with another; watching the darker area, the line of lysis that declared the presence of a bacteriophage. Gradually, I could classify my strains. Define their properties. Identify those in which ultraviolet radiation induced the production of phage and those that irradiation only killed. Bit by bit, I was creating a little universe over which I reigned. I was fabricating a "system" within which I reveled to my heart's delight, I played in total freedom; in short, I was concocting my experiments in total innocence.

The living world is one of complexity, the result of innumerable interactions among organisms, cells, molecules. In analyzing a problem, the biologist is constrained to focus on a fragment of reality, on a piece of the universe which he arbitrarily isolates to define certain of its parameters. In biology, any study thus begins with the choice of a "system." On this choice depend the experimenter's freedom to maneuver, the nature of the questions he is free to ask, and even, often, the type of answer he can obtain.

With my system of bacteriophages and of pyocyanic bacilli, I had in a few weeks learned certain characteristics of lysogeny. Within one species, ultraviolet radiation had induced the production of bacteriophages in certain strains but not in others. And in the bacteria-prophage association, the "inducibility" or the "noninducibility" characteristic was a property of the prophage, not of the bacteria. This was shown, for example, by the preparation and the study of bacteria harboring *two* types of prophage, one inducible and the other not. Irradiation brought about the production only of the inducible phage. Three months after my arrival in the laboratory, I was proud to publish my first note in the *Comptes Rendus de l'Académie des Sciences*, entitled "Induction of Lysis and of the Production of Bacteriophages in a Lysogenic *Pseudomonas pyocyanea*."

Biologists abhor abstraction. Let there appear a ghost, a creature of pure reason, and they will immediately form a visual representation of it. For us, the prophage was a particle, a granule contained in the lysogenic bacterium. Thus, we were able to discuss it, draw it on the blackboard or on the corner of a tablecloth in a restaurant. Questions then arose of themselves. What is a prophage made of? How many are there per bacterium? Where is it located? Does it float freely in the cytoplasm, or is it, on the contrary, attached to some structure of the bacterial cell? What is the mechanism that makes it inactive, incapable of producing viruses? How does it happen that ultraviolet radiation reactivates the prophage? All questions there was no way to answer when they took this form. To make them accessible to experimentation, the questions had to be changed, broken down into parts.

What extraordinary material this bacteriophage was! What speed of manipulation it allowed! In the morning, starting cultures on their way, preparation of the mediums. After lunch, as soon as the lab was freed, carrying out the experiment proper. Finding out the results the next morning. That was the most exciting moment. As exciting as listening to the radio in a time of crisis or seeing the denouement of a crime film. Every morning, upon awakening, I knew that something was going to happen.

Every morning, I hurried to the lab to find out what information the cultures of the night before had yielded. Hardly had the results been checked than one had to draw conclusions from them, to prepare what would come next. Which, most often, took place in the public square: that is, in the corridor. The corridor was in effect the obligatory passage, the general meeting place, the agora. Everyone came there to get some item from a cupboard, or to pass round some new journal article; to relate successes or failures, to ask about the others' results; to seek advice or a chemical formula, to talk over some theory, or simply to chat. The minute they were obtained, the results of an experiment were immediately transmitted to the people in the corridor. There, they were dissected, cut up, chopped fine by one person or another. By Louis Siminovitch or Elie Wollman. Or by Monod coming from one end of the corridor to see what was going on. If the experiment stood up against this trial, that was proof of its soundness. One could then proceed to the next. To establish a protocol and put it to work directly lunch was over. To thus prepare treats for the future, an exciting morning for the next day, with a new harvest of results. Little by little, I became addicted. Doing experiments turned into a mania, a drug I could not do without. Al Hershey, one of the most brilliant American specialists on bacteriophage, said that, for a biologist, happiness consists in working up a very complex experiment and then repeating it every day, modifying only one detail.

Certainly there also came days of depression. Days when nothing worked; when the bacteria refused to multiply; when the mediums proved to be contaminated; when the results made no sense. Difficult, then, not to think oneself stupid, incapable of ever doing research, good only for hawking men's ties on a street corner. But most of the time, this discouragement did not last long. Very quickly, in fact, I had learned that to avoid falling into the abyss with each failed experiment, it helped to have several other irons in the fire. To have several themes in mind at once, so that a failure here was balanced by a success there; so that one area at least was always full of good news. Always one element to nourish the waiting and the hope. But for my fear of being incapable of playing a role in science, I found myself at ease in this world of dream and of doubt. A world in which one endlessly played at setting up a fragment of the universe which the experiment would rudely correct. I could quietly surrender to my manias. Cultivate the *idée fixe*. Wallow in it at leisure. Prepare every day my dose of stimulant for the next. Start over again without ever getting winded. The endless, desperate pursuit of a goal that retreated as one approached it. Anxiety, the need for living in the future eventually served as a motor. In this world, even my defects found a function.

And then we felt the earth trembling in biology. Unquestionably, important events were brewing. Unquestionably, the laboratory would take part in them. I had the exact, powerful sense that the world was turning, that history would be made here. That I could be, however minor my role, a real participant, be right where the action was. Just as in London in 1940, after joining the Free French and hearing de Gaulle, I had the sense, in this laboratory, of having found the right address.

In one of my English lessons, Francis Ryan, my neighbor in the lab, tells me a story. Toward the end of the war, M., a German biologist, discovers an intriguing phenomenon: a protozoan with two sexes that send each other, through the culture medium, sexual signals favoring conjugation. The world of biology is interested. M. receives invitations from all over to colloquia and seminars to present his research results. And the story, which has begun in simplicity, develops in harmony. Each year a new stone is added to the edifice. First, M. describes the basic structure; the system; the way of demonstrating the phenomenon, of making it quantitative, of measuring it. The next year, there comes the finding of a hormone that is produced by females and attracts males. The following year, M. demonstrates the genetic mechanism that underlies the phenomenon and allows only females to produce the hormone. Then comes the isolation of the hormone, its gradual purification which leads to its chemical formula. The hormone in question is a steroid, little different from certain sexual hormones in mammals. Biologists are more and more fascinated. Especially those working with micro-organisms. Francis Ryan invites M. to spend a year in his laboratory at Columbia University in New York. M. arrives, armed with his strains. He sets to work. Very quickly, the affair takes a bizarre turn. Only M. is capable of repeating the experiments. Also, there must be no one around him. In a few days, Ryan solves the mystery. Not a word of truth in the whole story! M. has made it up from beginning to end. For me, the most astonishing thing is that the hoax went on so long. Perhaps because M. was almost alone in studying this organism.

After many hesitations, I finally registered at the Sorbonne's college of sciences. Once again, it was Herbert Marcovich who had convinced me. Sooner or later, he explained, a doctorate in science would become necessary. So I would need an undergraduate degree. Which a doctor in medicine could easily obtain: just two certificates, instead of the four needed by a student with only a lycée diploma. I selected general physiology and biochemistry.

It had taken a lot to get me to this point: at thirty, with a wife and child, after the campaigns, after the wounds! Going back into the classroom, learning more lessons, taking exams, was intolerable. I was in a cold fury. And on the first day, when I found myself among a gang of boys and girls who were barely eighteen, I nearly told the university to go to hell. My classmates knew each other. They were delighted to see each other again. They joked among themselves. Their every gesture, their every laugh spoke of the complicity of age, of a generation from which I felt excluded. For them, I was the old man. I was coming from another planet. They were not interested in me. And when they looked at me, it was with mistrust, without speaking to me. I, too, at their age, when a first-year medical student, would probably have done the same, stared at such a quaint newcomer. I felt further isolated by the importance they attached to the university, to lectures, to professors. What was for them the year's chief occupation was for me merely an ancillary activity, a sort of distasteful purge. I was determined to do the absolute minimum; just enough to take the examination.

The minimum was first of all lab work. Attendance in the lab, unlike lectures, was mandatory. The work there was done in groups of two. To begin with, one had to find a lab partner. Of course, all the young people immediately paired off. There remained a few isolates, the atypical ones, the oddballs lost in the crowd. Fate assigned me a thin girl with steel-rimmed glasses, a chicken all skin and bones, a frightened mouse who talked with her eyes fixed on the tips of her shoes. It took me three months to learn that her Polish-born parents had been deported. Suzanne herself had been taken in by farmers in the center of France where she had lived for some time. On returning to Paris to study, she moved in with an elderly aunt. Her ambition was to work in a pharmaceutical laboratory where some cousin of hers was employed. Suzanne had one great virtue: she went to every lecture, and took down every word that was uttered. When she got home, she copied out each lecture word for word, underlining the headings and subheadings in red, blue, or green. After which, she had no objection to lending me her notebooks. Thanks to Suzanne, I needed to attend only the lectures that interested me while having access to the rest that I needed for the exams. But a lot of time was lost running back and forth between the Pasteur Institute and the Sorbonne! Happily, this situation lasted only a year.

My brief stay at the college of sciences revived an impression I had once had at the lycée and then again in medical school: an impression of compartmentalization. There was neither connection nor synthesis between the disciplines, between the certificates, often even between courses leading to the same certificate. To prepare for an undergraduate degree was like touring an archipelago where, on each island, an archpriest preached strictly for his own chapel. At the same time, I discovered how exciting, even provocative, teaching can become when it involves not knowledge that was acquired long ago and has already fossilized, but science that is still uncertain, incomplete, developing. Most often, a lecture became exciting, stimulated the desire to know more of a subject, or even to work in a field, only when the professor was himself engaged in research on it; so far as what he was speaking of was his life, his passion, his daily struggle. A situation, alas, too rare. The physiologists, for example, exhaled a boredom so profound that it seemed even to affect them. Obviously they were repeating, unaltered, a story learned years ago, a story that did not directly concern them. They spoke without pleasure. One listened without interest. Fortunately, nothing discouraged Suzanne. She swallowed it all; got it all down on paper; copied it; underlined it; recopied it; and obligingly passed on to me the physiology lectures.

Biochemistry was another thing entirely. Two out of three lectures were worth it, not just because of the subject, but because of the professors' ardor. Neither thermodynamics nor the kinetics of enzymatic reactions were reputed to be grippingly amusing. Their interest came from their being on the border between physics and biology. The two subjects recapitulate, in a way, certain constraints that the laws of physics impose on the activities of living organisms. A fairly dry subject. But René Wurmser displayed such ingenuity in explaining the formulas of physical chemistry, described them with such ardor, and illustrated them with such felicitous examples from his own research, that his course came to resemble a detective story. Nevertheless, the teaching that was most valued, as being the most deeply immersed in current research, was that of Claude Fromageot on the structure and properties of proteins. A field in great ferment, for many reasons: because an ever-greater role was being ascribed to proteins in the function as well as in the structures of living things; because the development of physical techniques— the ultracentrifuge, chromatography, X-ray crystallography— provided new ways to study these giant molecules; because, for the first time, the chemical structure of a protein, insulin, had just been entirely elucidated by Fred Sanger at Cambridge University. To universal amazement, behind the enormous architecture of the molecule lay the linear chain of a polymer. Its extreme complexity in three dimensions rested on an extreme simplicity in one dimension. The decoding of this structure meant a transformation in the way of representing proteins, of studying them, of asking questions about them. A new world was opening up.

Fromageot talked about this change and these new ideas with a precision and an enthusiasm stimulated by his own work. He himself was seeking to define the chemical structure of another protein: lysozyme. His lectures, like the lab work he had us do, were designed to familiarize us with the ideas, the methods, and the techniques of a science in the process of evolving. The journal articles he cited or had us discuss had been published a mere two or three years before. Each week, the lectures dealt more and more with the questions of the present. Each week, he grappled with still-unanswered questions. How could one not wait impatiently for the result? How not be on the lookout for developments that were already under way? Working on the certificate in biochemistry, I discovered not only the excitement of learning a new field, but also how to conduct a course: the need for teaching and research to be closely connected.

Each morning, it was with the same pleasure that I went on foot to the Pasteur Institute. I liked to walk the streets of Paris before the city was completely awake, among the groups of children who, with their satchels on their backs, dawdled on their way to school. Perhaps this morning walk made real the dream I had brooded over in Africa when, deprived too long of France and of Paris, I hardly dared hope ever to return. We lived near the Carrefour Montparnasse. I went along the boulevard Montparnasse past the old Montparnasse Station, in this old neighborhood of low houses, today replaced by tall buildings. Then the boulevard de Vaugirard. The boulevard Pasteur. The rue du

Docteur-Roux. The two wrought-iron grilles painted black on either side of the street. In the 1950s, the Pasteur Institute still had much open space on its grounds, which had the look of a public park picked out here and there with stone or brick pavilions. Lawns. Flowers. Chestnut trees. A greenhouse filled with tropical plants. I arrived before most of the other scientists, while the laboratories were still being cleaned. The tile floors were scrubbed with a disinfectant perfumed with citronella, which reminded me of the circus of my childhood.

Each morning also, it was a delight to arrive in "my" laboratory. To have "my" place in a team. To be part, finally, of an institution, one of the most prestigious. An institution where, for more than half a century, discoveries had been made that had transformed medicine. I was flooded with both the pride of belonging to the line of scientists sprung from Pasteur and the fear of not proving equal to the height of my ambition. With the idea of trying, as it were, to drive a car without a driver's license; of insinuating myself into this place like a housebreaker. A house that, as I got to know it, appeared a little strange, a little unusual in certain respects. A curious mixture of excellent science and *laissez-faire*, of boldness and routine, of paternalism and incompetence. And also a highly flexible organization, which differentiated the Institute from public establishments like the universities or the CNRS, and helped it avoid stagnation and the burden of bureaucracy. In all, a somewhat marginal place, well suited to a marginal type like me.

In science, the great man is, first of all, the one who knows how to spot the right problems at the right moment, while there is a chance of solving them. He is also the one who knows how to surround himself with the right collaborators, to find among his pupils those capable of becoming his successors and of developing the theories he has set forth, the disciplines he has established. What was exceptional about Pasteur is that, having founded a science and a new medicine, he had helped his collaborators off to a good start. He had fitted out a workplace exactly arranged to allow them to exploit his ideas. The Institute, founded to apply the "treatment of rabies" perfected by Pasteur and his assistants, had been built with public donations. In the enthusiasm for Pasteur's work, people throughout the world had contributed, the poor and the rich, the bourgeoisie and the nobility, the czar of all the Russias and the emperor of Brazil and the sultan of Morocco. Pasteur had bought land that ran along what was then the rue Dutot. He had had constructed an initial building for the so-called "Microbic" laboratories. Later, more land was purchased on the other side of the street. After Pasteur's death a second building was erected, and named "Chemistry," in symmetry with the first. Little by little, still thanks to the generosity of the public, to gifts, to bequests, the grounds had been enlarged up to the rue Falguière on one side, the rue de Vaugirard on the other. New buildings were raised: a hospital for infectious diseases, with two pavilions and an outpatient clinic on the rue de Vaugirard; a pavilion near the rue Falguière for research on tropical diseases; then a building for tuberculosis. Thus, not only were the succession secured and the Pastorian disciplines extended, dividing themselves into various branches, but everything unfolded as the founder had planned. Everything functioned the way he had wanted, developing in the directions he himself had indicated.

I learned this story bit by bit, in snatches. At lunch or in conversations with the other Pastorians of the attic. My chief informant was Elie Wollman. For him, the Pasteur Institute was merged with France. He had been born in it, so to speak. He had lived most of his life in it. His parents had worked there until they were deported. His godfather had been the Russian Elie Metchnikoff, an immense scientist, the discoverer of phagocytosis and of cellular immunity, who had moved to Paris to do research with Pasteur. Having grown up in the Pastorian harem, Elie Wollman knew all its stories, all its secrets. His devotion to it, however, did not prevent him from vigorously speaking his critical, even acerbic, mind. Respectful of institutions, fond of traditions, Elie loved to condemn individuals. Few colleagues found favor in his eyes. His raillery spared neither the head nor the administration of the

Pasteur Institute, who "weren't what they used to be." Thanks to Elie, I began to grasp the structure and functioning of the Institute. He explained the advantages of a private center, though one officially recognized as of public utility, benefiting from the good will of the government and free to engage whomever it thought best. He praised the publications and the teaching that the Institute devoted to the different aspects of microbiology and immunology. He traced the branching out of the Institute to every continent, the gradual weaving of a network of Pasteur Institutes in the former French colonies, in the Far East, in Africa, in Madagascar, in Guyana. . . . He described the activities of doctors abroad: the struggle against parasites; against malaria, with Alphonse Laveran; against typhus, with Charles Nicolle. The prestige that this enormous work brought to France. In South America, for example, where Elie had lived for several years as a child when his father directed the Institute of Microbiology in Santiago. Pasteur's vision of the relation between research and industry. The twofold structure he had given his institute, where research fed ideas to industry which in return provided funds for research. A system that had worked brilliantly until the last war, while there was an important market for the Institute's specialty, serums and vaccines. The decline of this market following the arrival of sulfa drugs and, above all, of antibiotics. The bungling of the Institute whose chemists had developed sulfa drugs, but which left their commercial exploitation to private industry without receiving any of the profits. The Institute's inability to handle the change to antibiotics. The consequent financial difficulties in the early 1950s, which, according to Elie, might well worsen in the years to come.

Each year, at the end of September, everyone who worked at the Institute gathered to commemorate the death of its founder. "You should go, once, just to see," André Lwoff told me, with a smile. At the appointed hour, I followed the crowd of people emerging from their laboratories and going to the garden toward the Institute's oldest building, where Pasteur was buried. When I arrived, the hall was already filled: young and old, department heads and cleaning ladies rubbing elbows, wearing smocks or city clothes. All were murmuring, greeting each other, gossiping in low voices. On entering, I ran in to General-Doctor Marcel Vaucel, who gave me a friendly nod. The former director of the Free French Africa Health Service, the successor to General Sicé, he had just been appointed director of the Pasteur Institutes overseas. Apart from him, I knew hardly a soul. I stayed at the back of the hall, behind the crowd. Standing on tiptoe, I looked for familiar faces. I had, on the one hand, a mental list of faces seen here and there, and, on the other, a list of names heard in conversation. But, a neophyte still, I could not put the two together.

Suddenly I found myself next to Pierre Schaeffer, also an assistant in Lwoff's lab. He was the son of a professor of physiology at the Sorbonne. Having entered the Institute some ten years earlier, he had a good knowledge of the scene. Tall and blond, with very blue eyes and a handsome face often lit with a broad smile, Pierre pointed out Pastorians, describing each with a caustic, scathing humor. He indicated the luminaries of the Institute. First, the eldest, who still cultivated the historic look of Pasteur and his pupils as they appeared, with white goatee and black skullcap, in the portraits in the library. For example, Monsieur Camille Guérin—the G of BCG,* the antituberculosis vaccine—a short, slight man often seen strolling through the gardens. Then came the following generation, the men who had known those who had known Pasteur, and today held a bit of knowledge and power. "The important thing," Pierre explained, "is to be recognized as the champion in some field, to be considered indispensable in some specialty. Of course, you don't actually need to be the best. You simply have to be thought the best. So that you are identified with the field. So that you are the person they think of when they talk about it." And in the distance he pointed out a very tall, balding man with glasses: viruses. Next to him, a pink-

---

* Abbreviation for Bacillus Calmette Guérin, from the names of the two Pasteur Institute scientists who prepared the vaccine.

faced man with white hair: rickettsia. The short man over there with the tragic face: anaerobic bacteria. Another, very dignified, very red-faced: plague. Every germ had its representative, its defender, the sole expert entitled to speak on the subject. Then, the other specialists: the hospital doctors; the chemists who worked in liaison with industry; the veterinarians who developed serums for horses; the "colonial" doctors who maintained the affiliation with the Institute's overseas branches. There were those who shyly took refuge behind the others and those who wormed their way to the front row to be seen. And there were many other specialists, whom Pierre described with relish. The one who exercised the *droit du seigneur* over his female technicians. The one who came in at noon to pan-fry his steak on the gas burner before taking off for the day. The one to whom all France sent microbes that were hard to characterize, and who made his diagnosis by smell, sniffing at the culture tubes.

A sudden hush signaled the arrival of the dignitaries: the directors, and the board of trustees escorted by its chairman, a famous doctor who preserved chromosomes coming directly from the founder himself. Short, smiling, even jovial, he had, in Paris, directed the Resistance's health service. At the end of the war, he appeared in the subway with a false beard, but his leather satchel bore his initials in outsized letters. The director's brief address reminded the personnel of the virtues on which were founded "our house," its continuity and traditions.

Then, in silence, the descent into the crypt began, in Indian file, in hierarchical order: the director and board; council; then the department heads, the eldest first; the heads of laboratories, their collaborators; then the technicians and assistants; finally, the cleaning women and lab boys. Each went slowly down some steps before passing in front of the tomb. A surprising, neo-Byzantine mausoleum of marble, gold, and mosaics in vivid colors. With a cupola, columns of porphyry, and arched vaults. At the entrance, over the whole of the vault, mosaics depicted, in the manner of scenes from the life of Christ, those from the life of Pasteur: sheep grazing, chickens pecking, garlands of hops, mulberry trees, grapevines, representing the treatment of anthrax, chicken cholera, the diseases of beer, of the vine, of the silkworm. And at the summit, the supreme image, the struggle of a child with a furious dog, to glorify the most decisive battle, that against rabies. In the center, on the cupola's pendentives, four angels with outspread wings: three representing the theological virtues of Faith, Charity, and Hope; the fourth, judged fitting by turn-of-the-century scientism, representing Science.

This mausoleum out of Byzantium, these pious images, these golds, these marbles, the worship thus rendered, seemed out of keeping with the rest of the Institute, its style, its austerity. All this ostentation seemed pointless, for science as for the renown of Pasteur himself. The legend maintained by his family, then taken over by the Third Republic in accordance with the ideology of the time, had not merely glorified the scholar and his work but had deified the man, had made him into a saint and his life into a cartoon strip. Not a village in France lacks a rue Pasteur! An overpowering father figure in whom many Pastorians liked to see themselves reflected. Suddenly, there in the mausoleum, I recalled a scene with my grandfather. It was during the vacation in Le Mans, where he was commander of the army corps. I was seven or eight when I expressed my brand-new devotion to Napoleon, of whose exploits I had just learned. He had looked at me, smiling: "Admire, yes. Idolize, no. Neither gods nor men. Not gods, for they do not exist. Not men, for they are not gods." This formula had already haunted me during the war when, in the Free French forces, many of my comrades made a sort of cult of de Gaulle.

The tomb, covered by a block of dark granite, lay exactly under the cupola. Approaching it, each Pastorian paused for a moment, then walked slowly round it. In back, the line moved forward in fits and starts, shuffling in silence. On the walls to either side of the tomb were marble panels on which were carved, like Napoleon's victories in the Invalides, Pasteur's victories. Instead of the battles of Austerlitz, Jena, and Friedland, one read of molecular dissymmetry; fermentation; so-called spontaneous generation;

studies on wine; silkworm disease; studies on beer; virulent diseases; vaccines; prophylaxis against rabies. And even if one tried not to yield to the surrounding paternalism, even if one stifled all adulation, how could one not marvel at this unbroken series of triumphs, at this sure ability to deduce, from a theory, its applications or, on the contrary, to extract from the most theoretical problem, the most concrete aspects? How could one not admire this scientific odyssey, this way of vaulting from one domain to another, of going from chemistry and crystallography to the study of living things, from the diseases of beer to those of man?

Like Napoleon, Pasteur had fought many battles. There was, in fact, a military, a strategic side to the man. He had something of Napoleon in his way of always taking the initiative, of suddenly changing the terrain, of showing up where he was least expected, of suddenly concentrating his forces in a narrow sector to make the breakthrough, of exploiting his successes, of following up on the results and even of doing his own publicity or of bending others to his views. Like Napoleon's, Pasteur's art consisted in always joining the battle at the moment of his own choosing, at the place of his own choosing, on his own ground. And his ground was the laboratory; his weapons were experiments, protocols, the culture flasks. Whatever new domain he entered, whether he was interested in the grapevine or the silkworm, in chicken cholera or in rabies, Pasteur sought each time to transform the problem, to translate it into other terms, to open it to experiment. Today, we proceed no differently. What biologists do, above all, is to reformulate a lot of disparate problems so that they become accessible to laboratory experimentation. All their efforts aim at asking questions to be answered by experiments. It is with Pasteur and this strategy that began modern medicine and what is now called "public health." Without a doubt, Pasteur's saga was as stirring as Napoleon's!

When my turn came, I slowly walked around the tomb. I then went up the steps to come out of the crypt and leave the microbiology building. Outside, the sun of early autumn shone. On the trees, the leaves were beginning to turn brown. The line of Pastorians stretched into the garden, winding majestically toward the tomb of Emile Roux a few dozen yards away, on a lawn, in the shade of chestnut trees. My colleagues advanced in short steps, chatting in low voices and softly chuckling, like a class of old schoolchildren excited by the arrival of recess. Beyond the particularities of gait and demeanor, beyond the differences in age and origin, there was a family likeness between these men and women. Pasteur not only had founded new sciences and a new medicine; he not only had constructed a place of work; but, to populate it, he had also engendered an unknown species, an unprecedented type of investigator: the Pastorian. Recruited from just about everywhere in the world, the Pastorian sometimes came from afar, from Russia, from Central Europe, from the Middle East, or from America. Though trained in science or medicine, he or she remained most often on the fringe of official organizations and careers. A doctor with no patients, a pharmacist with no drugstore, a professor with no teaching duties, a chemist with no industry, his status was defined only by a style and, above all, by a place of work: he worked at the Pasteur Institute. Such a variety of talents gathered in one place to study the same material from every viewpoint: that is how a biology laboratory is organized today. There, too, in what is now called "interdisciplinary research," the Pasteur Institute initiated.

The line of Pastorians reached the grave of Roux. Suddenly the air was rent by horrible cries. Screams as heartrending as those of a mistreated child. It was the monkeys clinging to the bars of their cages in the animal house. Seeing humans, they clamored their indignation. In the 1950s there still stood, near the rue Falguière, the old building that had served as a stable for the horse-drawn trolley system. It housed the Institute's livestock: poultry, rabbits, goats, horses, monkeys. The crowing of the roosters, the clatter of horses' feet, the fine smell of manure still made this corner of the Institute feel like the country. Only the monkeys appeared to be disgruntled. It was said that one chimpanzee was so habituated to having his blood taken that, as soon as a white lab coat appeared, he would extend his arm through the bars of his cage.

If Pasteur's mausoleum had seemed garish, the tomb of Roux, a simple marble pedestal, seemed spare. Very much in the image of the man, who had lived the life of an ascetic, who had spent all his life in a room in the hospital, sleeping on an iron cot. A pupil and collaborator of Pasteur, he had developed the antidiphtheria serum that saved thousands of children from croup. The successor of Emile Duclaux, who himself was Pasteur's successor, Roux had for nearly thirty years ruled the Institute with an iron hand. With no family, waited on by the nurses of the hospital, he was without needs and did not imagine that one could live any other way. In his eyes, the honor of belonging to the house of Pasteur was beyond price and justified the starvation wages. "Man is so made," he said, "that he works well only in need." Countless stories circulated about him. One day, there came to see him a Pastorian grown gray on the job: "I have a wife and five children. I have worked here for close to thirty years, and I am still a mere assistant." Roux stared at him for a moment: "You want to be made a lab head? Well, then, so be it. But don't let anyone else know." It took the last war to soften the regime and to regularize careers at the Institute. For a long time after his death, the ghost of Emile Roux—"Monsieur Roux," as they said—continued to reign over the Institute.

Suddenly, I found myself before the directors and trustees of the Institute, standing, lined up, side by side. After walking past Roux's tomb, each of the Pastorians came in turn to file solemnly before them and to shake hands, as at a funeral one offers condolences to the family. When this ceremony was over, the groups reformed, walking back and forth in the garden. In the September light, faces took on the complacent look that people often have at family gatherings when some event, a marriage, a death, gives them a chance to get together, to gauge what each has become, what accumulated by way of luck or misfortune, what remains to him of life and property. Men and women slowly strolling, as on Sunday along the main street of a small provincial town. They smiled at one another or turned away. Gave each other furtive glances or ignored each other. As if they were trying to please some, to surprise them, or, on the contrary, to show others their contempt. Seeing this ballet, I wondered how many of them had reached the scientific goals they had once set themselves. How many had realized their dreams of glory. How many still dreamed or even still believed in their dreams and how many had, without saying a word, given up. It is easy enough to tell from one's face or behavior the actor who is booed, the financier who goes bankrupt, the shopkeeper who has lost his business. But a scientist who takes the wrong path or does not discover anything? A failure in science may for a long time keep up the illusion, even delude himself, carefully keep up his own passion. Once he has dug his hole somewhere, acquired certain habits, found a livelihood, however meager, what can he do but go on? How, then, does his life differ from the lives of other men who spend every day in office or store from nine in the morning to six in the evening, until they die? How many years does it take in biology before one makes a discovery of some importance? Ten years? Fifteen, perhaps? Within five years, however, one ought to know whether one is headed for success or disaster. One needs to be clear-sighted. Abruptly, I decided to give myself five years to find out whether I had a chance in the profession. If not, I would give it up.

One day an Israeli appeared in the lab. The first one I had met since the founding of the new nation. A bacteriologist from Jerusalem, he was interested in phages and lysogenic bacteria. He was also interested in the laboratory, in its occupants, notably the Jews who worked there. Tall, black-haired, blue-eyed, he spoke English in a commanding tone, with a slight accent whose origin I could not identify. That evening, I took him to a café for a drink. He told me about life in Israel. The departure of the English. The war against the Arabs. The difficulties of the young nation. And, above all, the enthusiasm, the ardor that, beyond its heterogeneity, motivated this small nation. "You must come," he told me insistently. "You must join us. A Jew's place is now with us, in Israel. And nowhere else." I was a little shocked. The idea had never crossed my mind. The birth of Israel had provoked in me neither great emotion nor a sense of victory but mixed feelings: a great deal of sympathy for the "Jewish people" who had overcome, at such cost, death and horror; but also distress at the conditions in which the Arabs had been forced to yield ground; the hope of seeing the victims of the Nazis settled there, finally, in a land of their own, but also the fear of a future that appeared heavy with threat. Sipping his beer, the Israeli described for me the determination of his countrymen, the young and the old, the sabras and the newcomers. He told about the kibbutzim; the struggle, foot by foot, against the desert; the first efforts at industrialization. "You must come," he repeated. "We need you. We need determined people. Specialists. You'll be more useful there than here. You'll have a more interesting life." But I did not feel involved. Not for a second did I identify with a builder, a pioneer reclaiming the desert. And then circumstances gave this young state a military and religious scent I found unappealing.

But the Israeli would not let me off. He played on the sense of solidarity with Israel that every Jew of the Diaspora must feel. On this indelible trace left in the deepest part of every Jewish soul by thousands of years of history and of persecution. On this sense that, despite their dispersion, all the Jews suffer a common fate; that when a Jewish community anywhere in the world is persecuted, all the others feel themselves to be affected, threatened. But this sense of kinship, which I did indeed feel for the Jews of Israel, was rather moral and intellectual than emotional. How could I not feel close to people who, having survived Dachau or Buchenwald, had lost in the crematories the illusion that they could be citizens like others in the nations where they lived? How could I not want to help the Jews of Central Europe who, despairing of an assimilation now judged impossible, endeavored to realize the old dream of building a nation of their own? But my solidarity with them did not mean becoming a citizen. I felt much too French. Passionately French. In culture, in language, in literature, in my family tradition, in the dreams that had haunted me all through the war. I reacted first as a Frenchman. I felt myself a Jew only afterward. And I saw no need to justify, to an Israeli or anyone else, this attachment to France and its culture. This attachment existed, full stop. It belonged to the deepest part of me. "The Jews can fulfill themselves only in Israel," continued the bacteriologist with a sort of fanaticism that was beginning to irritate me. We separated, rather dissatisfied with each other.

Agar plates. Colonies of bacteria. Plaques of phages. Thousands of plaques. Small, round, regular, with a cloudy bottom. All identical. Until the day this uniformity was disturbed by the sudden appearance of a small clear area, standing out like a scratch on the bacterial lawn. And, after re-isolation, this area gave birth to clear zones that also produced clear zones. My first mutant! A virus modified in one and only one characteristic. A change that had happened abruptly, in a single blow, and was transmitted to its descendants. Exactly as biology textbooks describe mutations and their properties. Exactly as in the drosophila fly or in man.

Yet what this mutant brought was not simply the pleasure of finding one of the rarest and most mysterious, as well as the most regular and reproducible, phenomena in the living world. It was also a tool permitting the study of an essential feature of lysogenic bacteria: what was called their "immunity"; that is, their resistance to infection by the phage they produce. A property necessary for the very existence of such bacteria, and lacking which they would be killed by their own product. But also a property particularly difficult to analyze without some means of distinguishing the phages used to infect the lysogenic bacteria from the phages spontaneously produced by them. Now this means was precisely what the mutant offered: the lysogenic bacteria producing the "cloudy" type of phage were resistant to infection by the "clear" type of phage. One could then analyze this immunity, showing that the "clear" phages attached themselves to the lysogenic bacteria producing "cloudy" phage; that they infected them without being able to multiply; that they remained in these bacteria as inert particles, gradually diluted by the multiplication of those bacteria. But the presence of these quiescent and invisible particles could at any time be demonstrated. All one had to do was to irradiate these bacteria to induce the development of the prophage. They then produced the two types of phage: cloudy, coming from the multiplication of the prophage; and clear, coming from the multiplication of the infecting phage. A surprising situation: not only did the prophage itself remain entirely quiescent during the multiplication of the lysogenic bacteria, but its mere presence was sufficient to neutralize the mutant virus, to keep it from multiplying, to reduce it to complete inactivity. Without in any way changing its properties, for one could reactivate it at will. What was the explanation for all this? The lively discussions in the corridor of the attic centered on two families of hypotheses: one brought into play a hypothetical constituent of bacteria; the other, a no less hypothetical product of the prophage. According to the first hypothesis, the prophage blocked a bacterial structure required for the phage to multiply. According to the second hypothesis, the prophage synthesized a compound that blocked one of the reactions permitting multiplication. Choosing between the two was rather a matter of taste. Each person gave his opinion. Minds with a biochemical turn preferred the second hypothesis. Those with a genetic bent tended to favor the first one. Lwoff and Monod liked the second; Wollman and I, the first. But the important thing was to find a way to distinguish between the two. An excellent exercise for the imagination. A good theme for *idées fixes*.

Thus, with time, the problems preoccupying me grew more refined, and the questions I formulated more precise. My activity seemed both to narrow and to deepen. When I had arrived in the lab, I was interested in themes of extreme generality: life, the cell, heredity. I wanted to understand the nature of viruses; to use them to establish the borderline between living and nonliving things. But in a few months, I had lowered my sights. My interest began to limit itself; my questions to become clearer. I started sharpening the contrast between the amateur and the professional in science; between the student and the confirmed scientist. I now contented myself with wondering about the nature of the prophage, about its structure and position in the cell, about the nature of the immunity it conferred on the host bacteria. But what I gained in precision, I lost in universality and comprehensibility. That made it hard to explain my work to laymen; notably to Lise, whom I so much wanted to share my excitement and my hopes and who so much wanted to share them. She had a keen mind and great curiosity. She had, however, been trained as a philosopher and had a taste for metaphysics. Any discussion of immunity repelled her. Any reflection about the prophage set her to wondering about the origin of life.

There were experiments designed to provide answers to very precise questions. But there were also fishing expeditions, trials for the hell of it, attempts "to see what happens if." Rather as in the war, when patrols were dispatched to "check out" the enemy. With sometimes a totally unforeseen result, which seemed to unmask a new phenomenon, to reveal unsuspected horizons. As if, suddenly, there was unveiled a still unknown aspect of reality. Hence an intense excitement. A mad hope. A sense of power, of mastering the world. I could not keep these overflowings of imagination to myself. I had to share them, to tell others about the mirage. Right away. Without waiting for confirmation. But the others tried to calm me down. Distrustful by nature as by profession, they refused to be impressed by what did not yet exist. They sent me to my lab bench. Except for André Lwoff, with his inexhaustible capacity for wonder, his enormous generosity. Every slightly odd result lit in his eye a bluer flame than usual. Always ready to welcome the peculiar, he immediately dreamed up a theory that might account for it. And if one expressed a doubt, one was accused of being a "man of little faith." Unfortunately, too often in this kind of situation, the replication of the experiment did not produce the same results.

The stupendous discovery burst like a bubble. It disintegrated into dust. One had to come down a peg or two. To fall back all the harder for having soared so high. To admit the inadmissible. To go down the corridor, crestfallen, making apologies. Perhaps this might be the source of certain frauds committed some years ago by students in American universities. Perhaps the student who one day brings in a dazzling result which his supervisor and the laboratory believe, perhaps this student begins to deny evidence to the contrary. To believe himself in the miracle, against all the odds, in spite of the impossibility of reproducing the initial observation. To be so convinced of its reality that he does not hesitate to give nature a nudge, to fudge the data, to change a number or two. In short, to fake things so as not to have to retract anything. To hold on to his status, his privileged position as researcher. But, then, what a lack of awareness! What childishness to think that, if the problem has the least interest, the fraud will pass unnoticed! That others will not soon, very soon, uncover the sham!

Very quickly I had caught on to the advantages of working not alone in my corner but in close interaction with the other workers in the attic. For efficiency, for mutual criticism; but also for pleasure, the dialogue prevailing over the monologue. Members of the group often combined their efforts for some experiments with limited objectives. For my part, I quickly committed myself to a series of collaborations: with André Lwoff, investigating compounds capable of inducing, like radiation, the development of the prophage in lysogenic bacteria; with Jacques Monod and his pupil Anna-Maria Torriani, comparing the effect of ultraviolet rays on various bacterial processes; with Elie Wollman and Louis Siminovitch, looking for other phages, other strains of lysogenic bacteria inducible by ultraviolet rays. There, a surprise; in several strains of colon bacilli, irradiation released the synthesis not of a phage but of a "colicin," a protein produced by one strain of bacterial offspring and killing other strains of the same species. All this was mere bagatelle, a diversion. The important thing was to understand the nature and properties of the prophage. To find the mechanism of immunity.

In the quantities of information reaching me in print or by word of mouth, I needed time to find my way. To know how to pick out the aspects of interest to me. During the first months, in my eagerness to get my feet on the ground, I tried to swallow everything, to register everything. By evening I was exhausted, my head empty, unable to recall what I had tried to grasp. Gradually, I learned to be selective. To take and to leave. Not to read the whole literature, but to know, if necessary, where to find useful articles. To prick up my ears only for the interesting parts of the countless lectures, speeches, talks, and discussions I attended. In addition to the lab seminars and Institute lectures, the Club of Cellular Physiology, founded by Jacques Monod, met once a month to hear a talk, followed by a discussion, ending with a dinner in a Latin Quarter restaurant. Once a month, the Lwoff lab group moved up to the dark, high-ceilinged library of the Institute of Physicochemical Biology, to meet with the laboratory teams of Eugène Aubel, René Wurmser, Edgar Lederer, and Ephrussi, who worked there. The sessions were led by Ephrussi and Monod, old cronies who shared a taste for intellectual terrorism. Boris Ephrussi was without doubt the most outstanding figure in French biology. Very tall, very thin, very upright, with a long knife-scarred face, he had the demeanor and style of a great lord. Leaving his native Russia at the age of fifteen, he had kept its accent; and his rasping voice, with its rolled r's, further accentuated the formidable aspect of the man. After several stays in the United States, notably in Thomas Hunt Morgan's laboratory, Ephrussi had, with George Beadle, been one of the first people to try to establish a relationship between genes and the chemical reactions that determine the eye color of the drosophila. Having recently turned to the study of yeast, he had opened up a new chapter in genetics with his analysis of mitochondria. After the war, Ephrussi had become the first professor of genetics in France. For it took until the the middle of this century to create a chair of genetics in France! A brilliant and critical mind, loving great syntheses and unexpected connections, Ephrussi was an extraordinary lecturer and conversationalist. An excellent actor, too. When he wanted to seduce, he knew how to deploy the sweetness of his Slavic charm and tell stories all night long. He also knew how utterly to ignore people of no interest to him. "For years, I remained in the blind spot of his eye," Elie Wollman said of him. Very domineering, Ephrussi ruled his laboratory and his students with an iron hand. He did not hesitate to throw down the sink an experiment one of his students had taken the liberty of performing without asking his opinion. Which did not prevent him, some minutes later, of showing himself, toward this same student, the most mild of interlocutors. He could in a few minutes go from anger to sweetness, from reasoning to joking, from exaltation to melancholy. There was in him something of Ivan Karamazov.

Many stars appeared at the Physiology Club in the early 1950s. J. B. S. Haldane, who had just emigrated to India, with his very young, talkative, and peevish wife; Linus Pauling and his molecular models; Guido Pontecorvo; Fritz Lipman; René Dubos; Michael Heidelberger; and many others. An impressive array. With an occasional feeling of surprise when neither a man's face nor his figure accorded with the idea one had formed of him from his writings. A sense of excitement very often, of plenitude, at the brilliance of a lecture or the success of some research. As at any great performance, whether athlete's, actor's, musician's: every time, in short, one is struck by inarguable perfection in some endeavor. A sense of oddness, even of obscenity in some instances, at the sight of some elderly gentleman going through all sorts of contortions to construct hypotheses, flinging himself about to find experimental recipes, using all the tricks of a game that, like the gestures of love, suddenly seemed to me reserved for the very young. As if science was too serious a business to be left to grown-ups.

The lectures and seminars featured a wide variety of subjects within biology. It did not take long to discover that certain aspects of it bored me no end. Pure biochemistry, for example. In the early 1950s, the fashion was for the study of "intermediary metabolism," those thousands of reactions; of diverging, converging, or cyclic paths by which the constituents of the cell are elaborated. For nearly a century, biochemists sought to give minutely detailed descriptions of the flux of matter and energy going through the cell. They had gradually brought to light a network of interactions whose complexity never ceased to proliferate. I, however, did not succeed in finding the slightest erotic appeal in the compounds in C3, C4, or C6, which transformed into one another, in the alcohols that became acids, in the oxidations or in the reductions responsible for respiration and the recruitment of energy-rich bonds. It seemed to me that the main principles of this field had already been defined; that it remained only to polish up their details; that in a hundred years there would still be reactions to pin down, compounds to unearth. What fascinated me were the deeply hidden elements that underlie the form and functioning of living things. Constructions of pure reason, the genes, the particles no one had ever seen, but whose existence had to be admitted. The abstract structures which one endeavored, by indirect methods, to provide with a content and a mode of action. What excited me was to look for the true agents responsible. To chase after the murderers, as in a detective story.

To think in terms of genetics and not biochemistry suited an ignoramus like me. For the biochemistry of metabolism could not be invented: one had to learn formulas and reactions. Genetics, on the other hand, constituted a logical system that worked like an exact science, a combinatorial system of "factors" and "characters." From the visible, the "phenotype," one inferred the invisible, the "genotype." These abstractions required no reference to their origin and composition. Thanks to the discontinuity deliberately introduced into the discrimination of characters, it was enough to count, in each generation, the individual organisms belonging to each of the possible classes. This whole science was thus based on the existence of observable units, the individual organisms, which played in genetics the role molecules play in chemistry. Genetic analysis could be directly applied to the study of certain aspects of lysogeny; to the position of the prophage in the bacterial cell. In undertaking this study, Elie Wollman used bacterial conjugation, discovered in a certain strain of colon bacillus a few years earlier in the United States by Joshua Lederberg and Edward Tatum. By mixing two mutants, each incapable of synthesizing certain metabolites needed for growth, one obtained recombinants capable of producing all these metabolites. These recombinants were rare, however, and the mechanism of their formation was unknown. Nevertheless, the system allowed one to tackle certain questions. To find out, for example, whether the prophage was located in the cytoplasm of the bacterium or was, on the contrary, connected with its genetic material. Not long before, Josh Lederberg's wife, Esther, had found that the colon bacillus, whose conjugation had been observed, was lysogenic; that it harbored a prophage called "lambda." In the mating of lysogenic and nonlysogenic strains, Wollman analyzed the behavior of these characters among the recombinants. Elie's first results, like those of the Lederbergs, seemed to indicate a connection of the lysogenic characteristic with certain genetic factors of the bacterium. Unfortunately, the data from other matings did not square with this interpretation. Uncertainties about the mechanism of conjugation prevented Elie, who was very cautious in his conclusions, from inferring from his experiments a nuclear localization of the prophage.

My preference for genetics over biochemistry was further enhanced by the presence of certain American researchers, who were spending a sabbatical year in the lab: notably those from the Phage Group, such as Seymour Benzer, Gunther Stent, Aaron Novick, and Cyrus Levinthal. All were physicists or physical chemists attracted by Max Delbrück to the biology of the phage. All of them, like Max Delbrück himself, were ignorant of biochemistry. All of them, like Max Delbrück, were interested in genetics, in which they saw the only specialization in biology whose logical structure permitted speculations somewhat similar to those in physics. These American colleagues taught me a great deal. Not just physical chemistry, thermodynamics, and the kinetics of reactions; but also, and especially, a certain way of reasoning, of treating problems, a way that originated in physics. Rigor in conceptualizing as well as in performing experiments. A methodical analysis of results that was pushed almost to its limits.

Seymour Benzer and I shared my laboratory for all of the year 1951–52. During the first months, there were few exchanges between us. We did not keep the same hours. I arrived at nine in the morning; he, around one in the afternoon. As he came in, he would throw out a resounding "Hi!" and then, after lunch, immerse himself in the inspection of his cultures. During the afternoon, he would belch once or twice. Around seven o'clock in the evening, I would bid him goodnight and leave him to his nocturnal experiments. Later, we became good friends. Of average height, balding, with a fleshy face and a stoutish body, Seymour hid behind an impassive mask much charm and warmth. With a mind that was both deliberate and penetrating, he had the habit of not responding immediately to questions put to him. Some days later, he would come back to see you with an analysis that went straight to the heart of the matter. Curious about everything, people and countries, slang and cuisine, cheese and music, he loved exotic foods. Every day, at lunch, he brought some unusual dish—cow's udder, bull's testicles, crocodile tail, filet of snake—which he had unearthed on the other side of Paris and which he simmered on his Bunsen burner. And, with all this, a pronounced taste for good jokes. When an American colleague, a football fan, was in Paris and wanted to know where he could find out about football in France, Seymour suggested he try André Lwoff. When the American asked Lwoff which match he would recommend for the following Sunday, the bewildered André at first took him for a madman before answering, "I have never, sir, set foot in a football stadium." When Jacques Monod was invited to lecture at the university where Seymour taught, the latter introduced him as follows: "This is Doctor Monod. He says that in his youth he hesitated between two careers: biologist or musician. Having played music with him, I can tell you that he was right to choose biology." Monod did not appreciate this one bit.

During the last weeks of his stay in Paris, Seymour and I did a series of experiments together. We irradiated bacteria during phage development and measured the effect of ultraviolet rays as a function of time: another way to study the behavior of particles and of abstract reactions without troubling oneself about biochemistry. And it was both an enrichment and a pleasure to see close up the working of such a fine and meticulous mind.

In the spring of 1952, a great joy descended on our little family, which expanded with the birth of Laurent and Odile. To the great surprise of Pierre who, having been promised a little brother or a little sister, got both. He stared wide-eyed at these new companions and rivals. Puny and fragile at first, the two babies were quick to get the upper hand and to gaze out upon the world: a very dominating look on the part of Odile, a more fearful one from Laurent.

Then came time for the first excursions. For the first colloquia. First at Oxford, where the Society for General Microbiology held a meeting on viruses around Easter 1952. André Lwoff was invited to present a paper on lysogeny. Escorting him was a contingent from the lab which included Seymour Benzer, Gunther Stent, and me. We stayed in a college whose students were away for the holiday. A solemn pile of old stones, with long chilly hallways and small rooms where the beds were on their last legs, the mattresses sagging to the floor. At Oxford I discovered the feverish atmosphere of colloquia, where what matters at least as much as the public talks are the encounters, contacts, gossip, word of mouth, the chance to learn who is investigating what, of gleaning information, of explaining what one is doing or at least what one wants to leak out and let people believe one is doing. This symposium was organized by two plant virologists, Frederik Bawden and Norman Pirie, old cronies who loved to play the buffoon, trading jokes and metaphysical aphorisms, all in a rapid, choppy English which left me in a cold sweat. For the laboratory group, the most eagerly awaited talk was that of Salvador Luria, one of the prime movers of the Phage Group. But, in the United States, this was the era of McCarthyism, of witch hunts; and because of a youthful flirtation with Marxism, Luria was denied a passport. He had, however, sent along the text of his paper which was circulated among us. This report reviewed the current state of knowledge about the phage. Most important, it took stock of the central question: which of the two constituents, protein or deoxyribonucleic acid, DNA, carries the genetic specificity? Six years before, Oswald Avery and his team had shown that DNA played a role in phenomena of transformation in the pneumo-coccus. But these experiments had not completely convinced the biological world, since DNA appeared to be formed by the monot-onous repetition of four chemical radicals; hence was an unin-teresting structure, too uniform to store the subtle complexity, the infinite variety required to determine heredity. Luria's report declared in favor of the protein nature of the genetic material in phage. The reason being that in the obscure period of viral re-production, when the infecting particle has disappeared and the daughter particles have not yet appeared, fragments of viral protein are detected first.

Since he could not come to Oxford, Luria had asked his pupil Jim Watson, then doing some work at Cambridge, to read this report. Himself convinced of the essential role of DNA, Jim had come to Europe upon completion of his doctoral thesis. First, to a biochemical laboratory with Herman Kalckar in Copenhagen. Then, after a few administrative struggles, to the X-ray crystal-lography laboratory directed by Lawrence Bragg in Cambridge, where Max Perutz and John Kendrew were analyzing the struc-ture of proteins. There, Jim joined forces with a fellow named Francis Crick. Both dreamed of determining the structure of

DNA. At the time, to a French student who had not yet been inside an American university or seen its denizens, Jim Watson was an amazing character. Tall, gawky, scraggly, he had an inimitable style. Inimitable in his dress: shirttails flying, knees in the air, socks down around his ankles. Inimitable in his bewildered manner, his mannerisms: his eyes always bulging, his mouth always open, he uttered short, choppy sentences punctuated by "Ah! Ah!" Inimitable also in his way of entering a room, cocking his head like a rooster looking for the finest hen, to locate the most important scientist present and charging over to his side. A surprising mixture of awkwardness and shrewdness. Of childishness in the things of life and of maturity in those of science. In the little world of the phage attending the colloquium, Jim created a sort of revolution when, instead of reading Luria's report, he brandished a letter he had just received from Al Hershey, another American prima donna of the phage chorus. It concerned a new finding. A neat, irrefutable experiment. The protein of the phage contains sulfur but no phosphorus, while DNA contains phosphorus but no sulfur. One may then specifically mark the protein with radioactive sulfur, the DNA with radioactive phosphorus, and, after infection, observe the fate of each tracer. Hershey and his colleague Martha Chase had in this way noted that the DNA of the phage penetrated into the bacterium during infection. The protein, on the other hand, remained stuck on the surface and could be detached by the force of friction produced in a Waring blender. Hence the unimpeachable conclusion: the phage was made of a sort of protein syringe that contained DNA and injected it into the bacterium during infection. DNA was sufficient to ensure the production of new viral particles. The protein merely served for the transport and protection of DNA. One could not but marvel at the hard simplicity, the dry solidity of such an experiment. Enough to make you forget intermediary metabolism and reconcile you to a certain kind of biochemistry. And, then, this result had a direct bearing on lysogeny. Now the prophage, too, could only be the phage DNA. A DNA arranged in such a way that it became incapable of multiplying. This in a reversible way, since irradiation restored all its faculties. Thus, little by little, the questions about lysogeny became more precise.

There was an abundance of colloquia in 1952. First, at the end of July, at the Sorbonne, the International Congress of Biochemistry, which mightily stepped up the Latin Quarter's density of discussion about intermediary metabolism. Immediately afterward, the First International Colloquium on the Bacteriophage, organized by André Lwoff. A week's meeting in Royaumont, which included everyone who counted for anything in the world of the phage. Notably, all my heroes, all those who had constructed the biology of the phage and whose work constituted my daily nourishment. Hence my surprise at having so often to replace an image I had gratuitously visualized with the face that I encountered. At having to exchange the tall, slender Italian Salvador Luria with long, black, brilliantined hair, whom I had imagined, for a Salvador Luria who was short, dumpy, with rather sparse hair. At swapping a rotund Germanic Professor Max Delbrück, pink and bald, for a tall, dry athlete with a thick head of hair and steel-rimmed glasses.

A formidable character, this Max Delbrück. The very conscience of the Phage Group, which he had brought together and which he directed with a blend of fantasy and firmness. Raised in the German university tradition, this physicist, a former disciple of Niels Bohr, had turned to biology to understand what a gene was, its structure, its replication. Delbrück's rigor, his frankness, his way of going to the heart of a problem were combined with his surprising youthfulness, of mind as of body. He liked to go camping, to take trips to the mountains or the desert. He loved jokes and gambling. But the game that for him mattered above all else was science: an open, direct science without secrets or any affectation of mystery, where the efforts of all were joined in mutual complement. He cared little about who was first to bring off a particular experiment. The essential thing was that it had been done. Only coherence and relevance mattered: the coherence of theories and of conceptions; the relevance of facts. Many feared his cutting judgments, about people as about their work. When a seminar bored him, he did not hesitate to leave the room or ostentatiously to open up a newspaper. In discussions about work, he would look his interlocutor straight in the eye. He spoke rather slowly and a bit haltingly, with a slight German accent. He would often halt in mid-sentence, his eyes wandering across the ceiling, his hand on his mouth, searching for the right word. Then the sentence took off at double speed. During the week at Royaumont, he sat very straight in the first row of the audience. The statue of the Commander.

A sparkling week amid the old stones and tall trees of the Cistercian abbey where Marguerite and André Lwoff seemed to be receiving their friends. A week of getting things in focus. DNA was growing ever more important. With a talk by Al Hershey, who had refined his experiments with the Waring blender, no one could doubt the role of DNA as the very substance of genes. The same problem remained: how to attribute such a function to what Delbrück called "so stupid a molecule." There was also a major discussion about lysogeny, with various talks originating in our lab. And also a talk by Jo Bertani, an Italian who was working in Luria's lab in the United States. I discovered, to my great vexation, that he was doing about the same thing as I. Like me, he was interested in the immunity of lysogenic bacteria. Like me, he infected these bacteria with mutant phages for the purpose of determining certain properties of the prophage and the immune mechanism. By all indications, we were going to be, we already were, competitors. Which annoyed me. Another theme of debate: the bacterial conjugation used by Elie Wollman to localize the prophage within the bacterium. For want of understanding its mechanism, many in the Phage Group tried to cast doubt on the very existence of the phenomenon of conjugation. On this subject, everything was based on the interminable articles of Josh Lederberg and on the hour-long sermons he delivered on the occasion of a colloquium. Certain characteristics of the colon bacillus seemed to be controlled by determinants that were linked; others not. Depending on one's taste, one could infer the existence of a single chromosome or of several. In addition, the results obtained often depended on the way the matings were carried out. Josh found some good explanation for each anomaly, but many people were incredulous and saw only caviling and hair splitting. Nevertheless, Elie managed to convince his people that even bacteria have the right to copulate.

The week at Royaumont was as fruitful as it was agreeable. It gave me a chance to get to know most of those who were working on subjects connected with mine. It taught me some of their enthusiasms, their idiosyncrasies, their intentions, and the hopes that carried them in one direction or another. Some of their ways of operating and reasoning as well. In short, I gained there a new view of the biological landscape, of this science that had just been born and was beginning to take flight. And then I had to give my first talk at an international meeting. Which seemed to me like having to take an examination before a jury presided over by Max Delbrück who, impassive in the front row, was regarding me fixedly through his glasses. To my great surprise, the stage fright that had gripped me before the fatal moment disappeared as I talked. It seemed to me that my report was well received. That I had passed the examination.

There remained for me one further test to face: to give a talk in English, which was becoming more and more the sole international language of science. This happened late in the summer of 1952, at a colloquium on the genetics of micro-organisms held in Pallanza on Lake Maggiore, in foothills of the Italian Alps. One evening after dinner, I was to give a talk on "colicins." The prospect of having to speak in English spoiled my otherwise highly agreeable stay. I had written my talk with care, but felt unsure both of certain expressions and of my accent. Very kindly, Jean Weiglé offered to help me. We had met in Royaumont and become friends. Jean occupied a place apart in the Phage Group. A physicist, he had, at thirty, been named holder of the physics chair at the University of Geneva. Convinced that physics is the business of young people, he had decided to resign at fifty. Which he did. On the very day he turned fifty, he left to go around the world. In California, he met Max Delbrück who converted him to phage. Since that time, Jean had worked on lysogeny, during the winter in Delbrück's lab in Pasadena and during the summer in Edouard Kellenberger's lab in Geneva. Short and broad of back, with crinkling eyes, Jean loved physical exertion, walking in the mountains, long swims. He talked about science with warmth, seeking the right word, the persuasive expression. There was a good deal of the poet in him. A gift for seeing in things signs others did not discern.

The day of my talk arrives. Before dinner, Jean accompanies me to my room to hear me rehearse it. I read my text. He has me cut certain passages, change a few expressions. He corrects my accent. After which, we go to join Lise and the others for dinner. Dinner over, I hurry to my room to reread my text one last time. No text. I cannot find my papers. I look everywhere. No papers. I turn the room upside down. Nothing. Lise comes to help me. Still nothing. It is getting late. I am eaten up with anxiety. How can I pull it off without a text? Panic. Two minutes to nine. I have to go to the meeting room. Confront all those faces waiting for me. Begin speaking. Without my text. A dry mouth. I swallow some water. Gradually the vise relaxes. I slowly but surely pick my way along from one word to another. And suddenly I see at the back of the room the gleeful face of Jean Weiglé. He is waiting for me at the exit. "Forgive me for playing that trick on you. But it was the best way to persuade you not to read a speech. You came through beautifully. A few mistakes in English don't matter." An unforgettable lesson!

There was an aspect of a ballet, of a traveling circus to these recurrent colloquia where the same people were periodically reunited to perform the same number. Actually, this was a club. A closed club where you had to be a member in order to have a role in the play presented on stage. Research on the phage, the genetics of bacteria, were the business of a small number of people. Some twenty to thirty persons scattered all over the world. Each forming a little world of imagination, with some correspondents who from Los Angeles to Rome, from Paris to

Tokyo, disputed with each other, railed at each other, helped each other. A sort of rugby team where the ball went from hand to hand. Only the countries of Eastern Europe remained on the sidelines, for Lysenko still reigned in Moscow. Membership in the club gave one the right to a series of privileges: hearing the latest news well before its publication; knowing who was doing what; having access to certain strains of bacteria and viruses or to certain rare substances; the benefit of incessant criticism that obviated blunders. In short, a perspective on the scene as a whole, which allowed one to see certain questions die and others be born; certain objects of research appear and others disappear. Thus, in the course of recent colloquia, we witnessed the ascension of DNA, its rise to the heights. Nevertheless, the first article by Watson and Crick on the structure of DNA, appearing in *Nature* in April 1953, had not electrified me or anyone else in the laboratory. I had only skimmed through this article. The crystallographic argument went over my head. It was only some weeks later, at the Cold Spring Harbor Symposium, organized that year by Max Delbrück, that I appreciated the virtues of the double helix.

Receiving Delbrück's invitation to give a talk on lysogeny at his symposium on viruses was like receiving my membership card to the club. And boarding the ship with André Lwoff, I felt as though I was setting out on a campaign to conquer America. What a shock was our arrival in New York! The excessiveness of the city, the noise, the skyscrapers, the heat, the jets of steam in the middle of the streets. With people and things churning in constant agitation as in a cauldron where the future was brewing. Then immediately, not thirty miles away, the contrast of the little laboratory at Cold Spring Harbor, an oasis of calm and peace on a bay of Long Island: with grass and the sea, small wooden houses set here and there among the trees, along a road leading to a beach; with no other noise than the cry of seagulls and the lapping of water on the beach. In summer this laboratory housed the Phage Group, who came there to do experiments and some practical teaching. It was famous for its symposium which, every year, surveyed some developing theme. For the colloquium of 1953, Delbrück had gathered virologists studying plant and animal viruses along with specialists on the phage, in an effort to extend to the viruses of higher organisms the methods and principles derived from the study of the phage.

How was it possible not to be overwhelmed, upon debarking from Europe, at the warm, good-natured, unconstrained, easygoing atmosphere that prevailed in such a scientific meeting in the United States? No constraints, no ceremonies. No hollow speeches, no weighty terms. Everyone sat where he felt like sitting, next to whom he pleased. Without barriers or hierarchies. But at the same time the scientific aspect of the colloquium was compact and vigorous. Unyielding. With tense, sometimes impassioned discussions. With what was unimaginable in Europe: young students who did not hesitate to challenge the official stars, to question snow-capped professors, even to put them in their places like simple buddies. All these boys and girls were ravenous and elbowing their way along. A sort of horde unleashed on science like a pack of greyhounds after a cardboard rabbit.

Many excellent talks at this symposium, of developments, new ideas. But the star turn was Jim Watson's description of the structure of DNA, which he had just worked out with Francis Crick. Some weeks earlier, on learning the details of this structure, Delbrück had immediately decided to add it to the colloquium program. And to ensure that no one would be unaware of it, he had had copied and distributed to all the participants the two notes by Watson and Crick that had appeared in *Nature*. His manner more dazed than ever, his shirttails flying in the wind, his legs bare, his nose in the air, his eyes wide, underscoring the importance of his words, Jim gave a detailed explanation of the structure of the DNA molecule; breaking into his talk with short exclamations the construction of atomic models to which he had devoted himself at Cambridge with Francis Crick; the arguments based on X-ray crystallography and biochemical analysis; the double helix itself, with its physical and chemical characteristics;

finally, the consequences for biology, the mechanisms that underlay the recognized properties of genetic material: the ability to replicate itself, to mutate, to determine the characters of the individual. For a moment, the room remained silent. There were a few questions. How, for example, during the replication of the double helix, could the two chains entwined around one another separate without breaking? But no criticism. No objections. This structure was of such simplicity, such perfection, such harmony, such beauty even, and biological advantages flowed from it with such rigor and clarity, that one could not believe it untrue. There might be details to modify, some further specifications to be made. But the principles, the two chains, the alignment of the bases, the complementarity of the two sequences, all this had the force of the necessary. All this could not be false. Even without understanding the details of the crystallographic analysis, even without an affinity for biochemistry, one could not but admire a structure that responded so well to the requirements of genetics. One of the oldest problems posed since antiquity by the living world, heredity, had just been resolved in the properties of a molecular species. The production of the same by the same, variation, the reassortment of characters in the thread of generations: all that flowed from the complementary distribution of some chemical radicals aligned along two chains. By all indications, it was a turning point in the study of living things. It heralded an exciting period in biology.

When Henri was born, he was all brown, as if bronzed by the sun, with very black hair and eyes. When he arrived at the house, in Lise's arms, wrapped in a white blanket with a sort of white hood that came down to his eyes, he looked like a little Bedouin. He was as beautiful as the three others. One blond, two brown-haired, one dark: a veritable experiment in genetics. The four little children coiled around Lise were the very image of life. Each birth, for me, had been a rebirth. In the evening I hurried home to rejoin this beautiful woman with these magnificent children crawling and squealing around her. Each time I rediscovered them with a savage joy and an infinite gentleness. It was like the return of spring, the leaves once again on the trees, the sun, flowers. It was like a revenge on the war, on death.

A morning in May 1954. The room for thesis defenses at the Sorbonne. On the benches, the members of my family, Lise, my father. Those of nearly the whole laboratory except, unfortunately, Marguerite and André Lwoff who have left for a term in the United States. Slightly nervous, I await the arrival of the members of the jury. They file in one by one. First, Boris Ephrussi, who is presiding. Claude Fromageot, the biochemist. Philippe L'Héritier, professor of genetics with Ephrussi; he is studying a virus housed in Drosophila in a relationship reminiscent of lysogenesis. I begin my act. The induction of the prophage through the exposure of bacteria to certain radiations or certain chemical compounds known to alter DNA. The immunity that paralyzes the particles related to the prophage. The hypotheses about the mechanism of this immunity. The nature and localization of the prophage. In rather complicated experiments, I have managed to count the prophages. There are few of them in a bacterium. Two or three. As many as chromosomes. That fact together with Elie Wollman's finding with conjugation leads one to think that the prophage is a genetic element that is added to the bacterial chromosome, hooking itself onto it. Ephrussi frowns. A chromosome is a stable, intangible structure, subject only to rare mutations. This is not a game of construction. One does not add pieces here to take pieces away there. Without which there would be no permanence of shapes and characters over generations. The members of the jury each go over their comments. Then they leave to deliberate. Then they come back to award me the rank of doctor of natural sciences (*docteur-ès-sciences naturelles*). That's it. The rites of initiation have ended. I have my diploma. I have, henceforth, every legal right to practice science.

VII

In THE ATTIC, I felt in my element, like a fish in an aquarium. But I had few illusions. I knew only too well that I had boarded a train already well under way. Despite my newly acquired doctorate, I still felt like a passenger without a ticket. I saw only one way to avoid the conductor: charge, head lowered. Attack on all fronts.

Why such frenzy? Why agitate oneself so? There is nothing inevitable about doing experiments, constructing theories, arranging and rearranging facts. By what necessity do men expend so much passion, take so much pleasure in eternally trying to explore the world, to interrogate it? To this question, people who love science respond: through curiosity, through the desire to appropriate nature, to improve the human condition. People who do not love science say: through ambition, through the will to power, the love of glory, or even greed. But that is not all. There are deeper reasons. There is the attempt, the temptation to understand a world that is veiled. The revolt against solitude. Against a reality that escapes you, is unaware of you, and without which there is no life. A metaphysical need for coherence and unity in a universe one seeks to possess but does not even manage to grasp. Nature is not mute. It eternally repeats the same notes which reach us from afar, muffled, with neither harmony nor melody. But we cannot do without melody. We have desperately sought it on the earth and in the sky before perceiving that no one will ever come to play for us the longed-for music. That it is up to us to strike the chords, to write the score, to bring forth the symphony, to give the sounds a form that, without us, they do not have. Such was, to my eyes, the function of science. Science meant for me the most elevating form of revolt against the incoherence of the universe. Man's most powerful means of competing with God; of tirelessly rebuilding the world while taking account of reality. Science manifested the passionate obstinacy of the human adventure in all its amplitude. Thus, being a member of this exceptional laboratory, working with these exceptional men, taking part in the new developments that were shaping up in biology, all these things stirred me up much as had being part of a Free French combat unit during the war. Once again, I felt, deeply rooted in myself, the sense of being where something was happening. And this, too, was a challenge. The opportunity to prove what I could do. But prove to whom? To my father? To my grandfather? To myself?

The thesis had marked the close of an era. Almost immediately, another era began, marked by a close and friendly collaboration with Elie Wollman.

How does one relate a piece of research work? How does one retrieve an *idée fixe*, a constant obsession? How does one re-create a thought centered on a tiny fragment of the universe, on a "system" one turns over and over to view from every angle? How, above all, does one recapture the sense of a maze with no way out, the incessant quest for a solution, without referring to what later proved to be *the* solution in all its dazzling obviousness. Of that life of worry and agitation there lingers most often only a cold, sad story, a sequence of results carefully organized to make logical what was scarcely so at the time. There also survive faces and words associated with certain events. Days, too, that have emerged from the grayness. Days lived with more force, more intensity; days that have remained on the level of consciousness. Like that morning in the summer of 1954 when, on arriving at the lab, I found my daily ration of bacterial colonies and results. A morning when I was expecting nothing exciting from the experiment Elie and I had set up the day before. From the first agar plate, however, from the first glance at the countless plaques of phages where I had not thought of finding them, I knew that a new era was beginning. That everything was going to change in the study of lysogeny as in that of microbial genetics.

There had been a new development in the field of bacterial conjugation since Elie's first experiments to localize the prophage. In several laboratories, those of Lederberg, Luca Cavalli-Sforza, and a newcomer, William Hayes in London, there had been noted the existence of sexual differentiation in the colon bacillus. Two types had been discovered: one behaving as a giver of genetic material, a kind of male; the other behaving as a receiver, a female. But there was disagreement about how to conceive of this conjugation. For Lederberg, the conjugation corresponded to a fusion of two bacteria, male and female; Hayes, on the other hand, postulated an asymmetry in the roles of male and female. For him, the male's contribution was limited to transmitting fragments of genetic material to the female. Until then, the extreme infrequency of recombination had not permitted an informed choice between these two interpretations. Recently, however, Cavalli-Sforza and Hayes had isolated new strains of males that produced high-frequency recombinants. Elie, who had spent a few days at Hayes's laboratory in London, had brought back a ''high-frequency'' strain. One could thus hope to succeed in specifying the localization of the prophage. With this strain, we again mated lysogenic with nonlysogenic bacteria; that is, bacteria with and without the prophage. Once again, the prophage was transmitted to recombinants by females, not by males. But we did not know why!

In the previous day's experiment, we had once again performed the series of matings between lysogenic and nonlysogenic bacteria. But we had added a new element: besides the recombinants, we had tracked the behavior of the prophage, convinced for other reasons that nothing would happen on that side. Hence my surprise at the sight of many unexpected plaques of phage. Hence too our satisfaction at finally finding some explanation of the mystery. For the virus multiplied each time the chromosome of a male carrying a prophage was transferred into a female with no prophage. Reciprocal mating produced nothing similar. Hence, finally, an explanation for what had hitherto been incomprehensible; if one did not find any prophage in the recombinants formed in the first mating, the reason was that, in this mating, the prophage developed and killed those recombinants in which it should have been found! A phenomenon immediately baptized "erotic induction" of the prophage; which for purposes of publication was to be changed to "zygotic induction."

It is rare that a single experiment touches on two fields of research at once. That it brings them closer, fuses them. With erotic induction, not only did conjugation make possible the analysis of lysogeny, but at the same time lysogeny proved to be an incomparable instrument for studying the mechanism of conjugation. On the one hand, erotic induction showed that genes transferred by conjugation are capable of expressing themselves in the absence of any genetic recombination: it was like a lamp that lights up to signal the success of copulation. On the other hand, the asymmetry observed between the two reciprocal matings provided an answer to the question raised several years earlier by the immunity of lysogenic bacteria. Among the recombinants formed in the two reciprocal matings, the only difference lay in the cytoplasm of the females. So it was the presence of some substance in the cytoplasm that blocked the expression of the prophage. It remained to find this substance and to determine how it worked.

All these facts were the result of close collaboration with Elie.

A collaboration that developed naturally. Smoothly. Without forcing. Elie and I had rather different but complementary natures. He was a night person; I, a morning one. He concentrated on a single problem; I always had three or four going at once. To advance, he always guarded his rear; I plunged ahead into the fog. He dropped a question only after he had exhausted it, reconfirming his results down to the last detail; once obtained, the results no longer interested me. In short, our personalities could either complement or confront each other. At the time, we never left each other's side. We spent our days together counting, pipetting, discussing. In the morning, as soon as the cultures were taken out of the incubator, and the colonies and the plaques were counted, we met to analyze the results, comment on them, prepare the day's experiment. After lunch, we performed the experiment. Often together. It was a matter of mixing cultures, series of different males and females, in various combinations and proportions. Of gently agitating them at 37 degrees centigrade. Of then taking samples at certain specific times and pouring them onto the agar plates that revealed either colonies of bacterial recombinants or phage plaques. Inexpensive experiments, for they used only easily available material. Experiments of great simplicity, with just a few minutes of commotion during which one had to avoid muddling everything up. But these matings of bacteria, this combinatorial system of couples resulted from a complex intellectual construction. They expressed our representation of the bacterial genetic material, its role, the way in which conjugation came about. Elie was an expert at this game. He combined good common sense and great acuity with an excellent way of analyzing the interaction between two bodies, an ability he had acquired during his stay at Max Delbrück's lab. The day ended in one or the other of our adjacent laboratories, with our constantly refining our representation of the bacterial cell, modifying it, imagining experiments to put it to the test. We looked like Abbott and Costello. Or Laurel and Hardy. Once, we even shared a technician assigned to us by André Lwoff. Monique G. was a sturdy, beautiful girl of twenty-four or twenty-five whom

Elie called "the saint." I never knew whether he had given her this name because she had to put up with the two of us, or because some mornings she arrived at work exhausted, her eyes betraying the effects of a busy night.

The end of 1954 brought changes in the management and the personnel of the attic. On the one hand, Jacques Monod left us, having, on the death of Michel Macheboeuf, been named head of the department of biochemistry. He and his group emigrated to the ground floor of the same building into spacious quarters with large, bright laboratories with high ceilings. From then on, it was there that the people from the attic and the ground floor took our common lunches, in a sort of glassed-in room adjoining the Monod laboratory's library. Which finally freed me of the daily torment of waiting for everyone to finish eating in the room where I worked. On the other hand, on their return from the United States, Marguerite and André Lwoff had changed their research topic and material. They gave up phages and bacteria to study the polio virus, which required new equipment and a re-shuffling of laboratories. Work on bacteria was carried on in the attic: on one side by Pierre Schaeffer, who was studying the transformation of bacteria by DNA; on the other, by the little group Elie Wollman and I had formed.

The "system" we had worked out, this combination of phage and bacterial conjugation, constituted a remarkable experimental tool. We now had to fiddle with it and amplify it, letting it ramify in every direction. The study of lysogeny had shown the necessity for genetic analysis in the study of any bacterial function. Conjugation seemed to provide the necessary tool for this analysis, insofar as it was possible to distinguish the main steps. This was no small matter. One could, with a microscope, observe the beginning of copulation, the pairing off, the formation of couples. One could see the end, the formation of recombinants. But between the two, there was no way to see anything whatever; to gain direct access to the intimate functioning of this sexuality. Everything became a matter of imagination and inference. A sort of detective story. We had to visualize the process: to try to disturb it by every means that came to mind; to watch for the resulting alterations in the normal course of things; and so to change our idea of the mechanism. Thus everything depended on the representation we formed of an invisible process and on the manner of its translation into visible effects. Again, no simple matter.

How, for example, could we analyze the way in which the chromosome from the male was transferred to the female? We pondered this for a long time. Until the day when Elie was visited by a strange idea. One could imagine the following scenario for conjugation: male and female pair off; after a certain period, a transfer of chromosome from male to female, the latter thus being promoted to the rank of zygote; genetic recombination between the two chromosomes in this zygote. If the two partners were abruptly separated *before* the chromosomal transfer, there ought not to be any recombinant. If they were separated *after* the transfer, then the recombinants ought to form. Hence Elie's suggestion: to mix males and females; then, some moments later, to separate the partners abruptly by placing the couples in a Waring blender. A sort of *coitus interruptus*. On my first trip to the United States, I had brought back a blender for Lise to make purées for the children. But Lise detested the machine and would not use it. So I had stored it in the laboratory where it might someday prove useful. And that day it did.

The next morning in the lab, a breathless arrival. A rush to the culture plates. And there, wonder of wonders! As expected, no recombinants at the start of the experiment. As expected, the sudden appearance of recombinants whose number increased over time. The surprise was that the two characters under study were behaving differently. One arrived fifteen minutes before the other. As if they had been localized on two independent structures, on two chromosomes transferred at different times. Hence, great excitement and animated discussions with Elie. Either this was an error of manipulation or we had found an important result.

That very day, we repeated the experiment. This time, however, with strains differing by five, not by two, characters. And the next morning, the incredible had happened. Each of the characters of the male appeared on time: one at ten minutes, another at fifteen, still another at eighteen, and so on. But, above all, they arrived one after another in the very order in which they were known to be arranged on the chromosome of the male. The forces of friction set off by the blender had not only separated the couples. They had also cut the chromosome coming from the male while it was slowly moving between the two mating bacteria. As if this chromosome were transferred to the female not all at once, but gradually, linearly, at a constant speed, beginning at one end. As if this chromosome were being gently sucked up by the female like a strand of spaghetti. Hence the name "spaghetti experiment" Jacques Monod gave to these manipulations in the blender. To Elie's great displeasure, for the name lacked nobility, in contrast to the elegance of the experiment.

Erotic induction and the spaghetti transfer modified the whole representation of conjugation. We could recognize each of the steps, measure their effectiveness, predict fairly accurately the results of a mating. This gave me the impression that we had invented a story. Created a sort of scenario which I at first found hard to believe. This seemed fragile, slightly unreal, as if constructed for the occasion. Nevertheless, this picture held up to experimental testing. It was completed bit by bit by new research, with the participation of the foreign researchers who had come to the attic to spend a sabbatical year. A study, by Tom Anderson, of bacterial couples using the electron microscope; a genetic study of immunity in the phage, with Dale Kaiser; the introduction of radioactive phosphorus into the chromosome of male bacteria, with Clarence Fuerst. For the first time in the history of biology, it became possible to establish a chromosomal map by three independent methods. First, a genetic method: one could measure the frequency of recombinants formed in the matings, and thus the probability of chromosome breaks between two mutations. Next a chemical method: one could charge the chromosome of the male with radioactive phosphorus, measure the ruptures occurring in the chromosome through phosphorus disintegrations, and then estimate the quantity of DNA separating the two mutations. Finally a physical method: one could determine the times of entry of the genes into the female bacteria. The map of the bacterial chromosome obtained by this method could be compared with that of the Paris-Marseille axis that could be established by determining the times of a train's passing particular stations when traveling at a constant speed.

The three or four years spent studying bacterial conjugation, erotic induction, the *coitus interruptus*, was a period of jubilation. A time of excitement and euphoria. But my memory of it is frozen. It has crystallized in articles and reviews, abstracts and lectures. It has lost its color, dried up in a story too often told, too often formulated. A story that has become so logical, so reasonable as to have lost all juice, no longer conveying the sound and the fury of the daily research. What gave it life has been swallowed up by time. Gone are the abortive trials, the failed experiments, the false starts, the misguided attempts. Forgotten are the fallacious arguments, the hesitations, the jabs of the sword in the water, the groundless joys, the spurts of rage against oneself or against others. Vanished are the hours spent counting the colonies, the anxieties, the uncertainties, the endless waiting. Everything has become smooth and polished. A fine story, very clear, with beginning, middle, and end. With well-oiled, well-articulated, well-arranged experiments, one following another, leading without fault, without hesitation, in seamless argumentation, to a well-established truth. The truth found in textbooks on genetics.

Occasionally other fragments of the past come back to light. Loom up, intact. Impressions. A warmth suddenly coming to my cheeks, for example, as I come upon an old photo of Jacques Monod, with his ironical little smile. Immediately I am back in his office, seated before him. A room in the middle of the hallway, on the ground floor. I have come down to tell him about a result that turned up that very morning. A still uncertain result. Obtained once only. But I have to talk about it, tell the story, share my excitement. To think, to forge ahead, I need to discuss. To try out ideas, to see them rebound. And no one plays this game better than Jacques. He listens to me. Looks at me. Holding his chin in one hand, digging into it with his finger. He asks a question. Rises. Goes to the blackboard. Sketches a diagram. Returns. Abruptly asks whether I have thought of doing a certain control experiment without which my result is worthless. I feel myself redden in confusion. I have forgotten this control. A faint ironic smile plays over Jacques's lips. The smile in the photo. I would like the earth to open up and swallow me.

Or again, recently, the sound of breaking glass, mixed with the warm sweetish smell of agar medium, lights again in my heart a rage at myself, at my stupidity, my clumsiness. I am sitting in my attic laboratory. Next to me, Martine Tallec, the technician with whom I am working, is filling agar plates for the afternoon's experiment. I am feverishly examining the stacks of plates from yesterday's experiment. For more than two weeks, I have been looking for a mutant, a certain type of "high-frequency male." In vain. I am not succeeding. And yet I am quite sure of myself. I am sure of the existence of such mutants. What distinguishes male bacteria from females is the presence, in the males only, of a small genetic element, a little fragment of DNA, the "sex factor." The high-frequency mutants must form when this sex factor, free in cytoplasm, hooks onto the bacterial chromosome which it can then pull, like a locomotive, during conjugation. Still no mutant. Nothing on this plate. Nothing again. Still nothing. Suddenly a plate covered with colonies. That's it. At last. In my impatience, an abrupt gesture. The stack of glass plates comes crashing down on the tiles. What a pathetic idiot I am! The mutant was there! I had only to scoop it off. Will I always be such a klutz? Martine, who has been slaving away with me looking for mutants, gazes at me, desolated. The only consolation: that plate did prove the existence of these mutants. All we have to do is start all over again.

Other circumstances evoke other scenes that suddenly rise up out of the past. But always mere scraps, fragments. Quickly borne away once again by the story ready-made, frozen, neatly tied up. With Martine, I isolated a good half-dozen high-frequency males. A surprise: they all differed from each other in the order in which they transferred their characters to the females. But there was a coherence behind these differences. As if it was always a matter of circular permutations. As if the bacterial chromosome was a closed structure, a circle that the sex factor had opened by inserting itself into it here or there. When I first spoke to Elie about a circular chromosome, he shrugged and plunged into his notes without responding. He was against it, which came as no surprise, for he was readily against. Against novelty. Against the unexpected. Judging my arguments insufficient, he began to isolate his own mutants. They behaved the way mine did. Elie was persuaded. Some months later, John Cairns isolated the chromosome of the colon bacillus: it was indeed a closed, circular structure.

There were many resemblances between sex factor and phage. Many common properties. These properties defined a new class of genetic unit. André Lwoff, who adored new words, decided we had to find a name. Hence, a meeting of the "terminology committee" created by Lwoff for such occasions. For the viruses, for example, when it came to light that, whatever their form, their chemistry, their properties, they were always constructed according to the same principles: a nucleic acid, RNA or DNA, enclosed in a shell formed by the assemblage of protein sub-units. The nucleic acid served in virus reproduction. The protein shell ensured the protection and the infectivity of the nucleic acid. The terminology committee had long pondered a name for this shell. In an old Greek dictionary, I had found the word *capsa*, meaning "container." Hence a series of trials: *capson, capsus, capsal, capsard, capsin, capsoid*. Just like the sessions of old when as a child I repeated ad nauseam variations on the same word. Finally, the committee unanimously adopted *capsid* to designate the protein-aceous envelope of the viruses, and *capsomeres* for the sub-units constituting the capsid. As for the genetic elements represented by the phage and the sex factor, they needed a word ending in *-some* ("body"), like *chromosome*. Hence, *parasome, catasome, anasome, périsome, métasome*. Finally, it was *episome* ("extra body") that carried the day. Thus named, things immediately took on a new reality. They existed!

The five years I had set to prove myself had flown by. I never ceased to wonder at my having been able to approach what had long seemed an unreachable goal. At finally having obtained what, for years on end, I had not dared to hope for. That one could live, travel, eat, and raise a family while spending the best part of one's time doing what one loves, that seemed like a miracle I still found hard to believe! Finding myself in the trappings of a scientist had an element of the unreal, the unattainable. I knew my shortcomings well enough not to overindulge in dreaming and to keep my illusions within bounds. I was not unaware of the luck I had had: in the research topic assigned to me; in the men who had welcomed me; even in my faults, which proved suited to this activity—impatience, obsession, the need to go quickly, the compulsion to live in the next day. Because of our work, bacterial conjugation had become a unique tool for the analysis of any bacterial function. Already, the results we had obtained no longer interested me. The only thing that mattered was what we were going to do with this tool.

Fall 1956. The University of Caen. One of the oldest cities of France: more than half a millennium. Razed by the bombings of the summer of 1944. Just rebuilt. For some years, the new amphitheaters have been the seat of intense agitation: all of a sudden, politics is taking an interest in science.

The relations between politicians and scientists in France have never been close. Except at the time of the Popular Front, which ushered in a Secretary for scientific research. Also the development by Jean Perrin of the Caisse Nationale des Recherches, soon to become the Centre National de la Recherche Scientifique

(CNRS), which saved French science from total disaster. By habit, the successive governments since the war have continued to name a Secretary for scientific research. These appointments have nearly always reflected political rather than scientific requirements: for the title holder has been designated not for his competence but for the votes he would bring to the government. It has been, in short, a question of employing someone in a ministerial post without giving him too much importance. And suddenly a politician speaks of scientific research and its crucial role for the nation. A politician comes and says that in this century neither power nor prosperity is possible without science and technology. And not just any politician. The only one since de Gaulle to display a statesman's stature and breadth: Pierre Mendès France, who wonders and worries. Who denounces the inadequacies of our higher education and research. Who rages against stagnation, opposition to progress, ossification, bad habits, fiefdoms, privileges. And finally calls for a vast colloquium to define areas of neglect and to specify remedies.

A large amphitheater. On the benches, a confused mass of parliamentarians and researchers, industrialists and professors, administrators and social scientists. A feverish atmosphere of criticism and hope. An atmosphere of work and fervor. Each speaks his piece, details his discontentments, his proposals. The administrative compartmentalization. The lack of funds. The mediocrity of careers. The perpetuation of old disciplines and the absence of new ones. The isolation of research and of industry. The insufficiency of good administrative personnel in higher education. A ten-year plan. The reform of medical studies. An increase in the number of science students. Suddenly Raymond Dedonder, a researcher at the Pasteur Institute, rises. He goes to the blackboard. He draws the coordinates of a graph. He turns around to make a prediction that is hard to dispute: people who were born in 1945 will be twenty years old in 1965! On the ordinate, he places the number of students expected; on the abscissa, the successive years. Then he plots the curve. The number of students grows and grows. "In five years," says Dedonder, "if something is not done, the universities will be in dire straits. At this point everything will explode." He indicates a point on the curve: 1968!

For a researcher it is not enough to do experiments, to obtain facts, from them to form a theory. He must also make these results known. Convince his colleagues of the importance of his work, of the value of his theory. In short, he must seize every opportunity to advertise, to hawk his merchandise. To expose it to public criticism and commentary. To jealousy as well. For "what makes the profession of research irksome," said André Lwoff, "are the discoveries of others!" So Elie and I turned into traveling salesmen. Notably in the summer of 1956, when we were selected to take part in two major colloquia in the United States. Elie was to give a talk on conjugation and lysogeny at the annual Cold Spring Harbor Symposium. I was to present a similar number a few weeks later at a colloquium on the chemical basis of heredity at Johns Hopkins University in Baltimore. At these colloquia we saw developing what was to be molecular biology: that is, the study of structures, functions, and biosynthesis of the two great biological polymers, nucleic acids and proteins. This biology was performed by a very small, very exclusive club; a sort of secret society to which belonged perhaps a dozen laboratories throughout the world. Some forty researchers who wrote to each other, telephoned, visited, exchanged strains and information, traveled to each other's labs to do experiments, met periodically here and there to keep abreast of the current state of the field.

Our model of conjugation had encountered no opposition. Among the participants at the Baltimore colloquium was Josh Lederberg, the discoverer, together with Ed Tatum, of bacterial conjugation, whose views on the question were very different from ours. Even he voiced no objection. The following year, however, at a colloquium of the Society for Experimental Biology in London in September 1957, I was the target of a violent attack. This was my first chance to present abroad the hypothesis of a

circular bacterial chromosome. A young and pretty British micro-
biologist, who was returning from a stay in Lederberg's lab, said
to me peevishly: "What a stupid idea! We don't need a chromosome that goes through contortions or around in circles to explain our results." Also attending this colloquium was another
Frenchman, a professor at the Sorbonne: the very one whom
Lwoff had challenged to a duel at the time of the study section of
the CNRS. He gave a talk of such inanity that the English were
smiling among themselves. I wanted to be a hundred miles away.
I asked why he had been invited. The secretary had got hold of
the wrong address: our man had been invited instead of some-
body with the same name from Brussels!

But the star turn at this London colloquium was the talk by
Francis Crick. We had come across the name of Francis Crick in
the notes in *Nature* on the structure of DNA. At the time, no one
in the attic had ever heard of him. Everyone, on the other hand,
knew Jim Watson and his uncommon personality. So Crick had
seemed like some sort of residual appendage to Watson. Some
months later, however, when Crick made an appearance at the
Pasteur Institute to give a seminar, it was immediately clear to all
that Francis was not simply Jim's appendage; that a character as
strong-minded, as mentally acute had no need of a coach. Tall,
florid, with long sideburns, Crick looked like the Englishman
seen in illustrations to nineteenth-century books about Phileas
Fogg or the English opium eater. He talked incessantly. With
evident pleasure and volubly, as if he was afraid he would not
have enough time to get everything out. Going over his demonstration again to be sure it was understood. Breaking up his
sentences with loud laughter. Setting off again with renewed
vigor at a speed I often had trouble keeping up with. A formidable intellectual machine, Francis Crick played a major role in the
development of molecular biology. He had no taste for experimentation, for manipulation. But no one contributed more than
he to the working out of the body of hypotheses that in the 1950s
and 1960s guided experiments and made it possible to foresee the
general outlines of what was to come.

At the London colloquium, Crick presented an overview of the synthesis of proteins, a particularly unsettled subject at the time. The theme of countless works and publications where the best were cheek by jowl with the worst. On this difficult subject, Crick was dazzling. He had the gift of going straight to the crux of the matter and of ignoring the rest. Of extracting, from the hodgepodge of the literature, the solid and the relevant, while rejecting the soft and the vague. His talk was based on hypotheses that, although so far not well supported by direct argument, helped one to grasp this highly complex problem. For example, the so-called sequence hypothesis, according to which the sequence of the bases in a segment of nucleic acid *suffices* to define the sequence of amino acids in the corresponding protein. And then, it took a sure footing, an acute sense of publicity, to baptize Central Dogma—that is to say, incontestable truth—a hypothesis that was unsupported by any serious argument, but that, restricting the limits of the possible, sharpened the field of research. According to the Central Dogma, the information defining the sequence could go from nucleic acids to protein, but never in the reverse direction. Once in the protein, it could never get back out: neither back to the nucleic acids nor to other proteins. If, then, it was necessary to find the details of the machine that enabled the nucleic sequence to translate into a protein sequence, it was pointless to look for another machine for a reverse translation. Such a machine did not exist. To go straight to the point, not to worry about details, at least to begin with: such struck me as the lesson to be drawn from Francis's talk.

And also not to quail at the boldness of a hypothesis. Contrary to what I had long believed, the process of experimental science does not consist in explaining the unknown by the known, as in certain mathematical proofs. It aims, on the contrary, to give an account of what is observed by the properties of what is imagined. To explain the visible by the invisible. And it is through the evolution of the invisible, through an appeal to new hidden structures, with hypothetical properties, that science proceeds. Lightning, for example, was long considered an expression of the wrath of Zeus before it was viewed as resulting from differences in electrical potential between the sky and the earth. Infectious diseases were thought to be the effect of evil spells cast over sick people before the role of microbes and viruses was invoked. In either case, to explain a phenomenon one had to regard it as the visible effect of some hidden cause, which was linked to the ensemble of forces thought to control the world. It was, thus, the properties of these hidden structures, of these invisible forces, on which the imagination could play by constructing its theories. Nothing then ought to stop this imagination as long as it had some way of being put to the test.

Sometimes, in the night, I wake up with a start. In a sweat. Gasping. Hardly daring to breathe. Immobilized by anxiety. Astounded to find myself in this bedroom. To have emerged from the nightmare. Not to find myself in the mountains of Tunisia, walking in the moonlight past the Nazi soldier's submachine gun. Or not to feel the weight of the corpse of my friend Vincent D. whom I am taking to the first-aid station. Or, again, of not lying in the American field hospital, with Lieutenant B. nearby, whimpering in agony. Slowly, the world falls back into place. The window. The ceiling. The bed. Lise's regular breathing. Nothing stirs in the room. In the distance a dog barks in the night. In my mouth remains the bitter taste of death; in my eyes, the profiles of those who were my friends. Who, one after the other, have died. I get up and slip into the children's room. They are asleep with their fists closed. Lying every which way. I draw a blanket over one. Put a foot under a sheet. Fascinated by this marvelous quartet, I stand for a while watching this desire for the future gleam in the dark. Hearing sing, like the libretto of some unfamiliar music, this promise of immortality. Only these children's faces can blot out the faces of the dead.

There is a style in science. As in art, in literature, in painting. Not just a way of looking at the world, but also of questioning it. A way of acting with regard to nature and of talking about it. Of concocting experiments, of executing them, of drawing conclusions, of formulating theories. Of supplying them with a shape from which a story may be drawn, whether spoken or written. There is an infinite variety of styles. Direct or convoluted. Concise or multifaceted. A workman's style or a cavalryman's. An eagle's or a mole's. A visionary's or a follower's. A great lord's or a shopkeeper's. A paranoiac's or a melancholic's.

Jacques Monod had a very particular, a very personal style. Both obstinate and flamboyant. A mixture of logic and passion. Of tenacity along a single track, and probing thrusts in every direction. This puritan was bold. This atheist concealed a believer. Haunted by a moral exigency, by the need to look for the truth of nature and to make it known. To persuade his colleagues, and even laymen, of it. More than confidence, he had faith in this nature, in its coherence, its unity. Hence the famous aphorism: what is true for the colon bacillus is true for the elephant. To analyze the functioning of bacteria was to study the human being. What a pleasure to work with a man for whom to live was to search. In every field. In every direction.

Our collaboration had become inevitable. With Elie Wollman, we had developed a tool that made possible genetic analysis of any function, any "system." For his part, Jacques had such a system in his quest for genetic analysis. In the fifteen years since getting his doctorate, he had been tirelessly digging away at the same furrow, using every new technique, every method, gradually constructing, by little touches, an ever tighter representation of the mechanisms that underlay this function. In the fall of 1957, he had arrived at the exact point where further progress required genetic analysis. Elie had no wish to take part in this. He withdrew into his tent to write his thesis. Jacques and I decided, therefore, to work together more closely on the study of his system: the metabolism of lactose, milk sugar, in the colon bacillus. The first series of experiments suggested itself. Conceptually simple, but difficult to perform technically. To use lactose, the colon bacillus employs the very stable enzyme galactosidase, whose activities are easy to measure, and a "permease," which sucks up the lactose from the medium into the bacterium. Monod had isolated mutants that were unable to synthesize either one or the other of these proteins. To check out the system, to regulate it, to control it, we decided to begin with the following experiment: to have a "high-frequency" male inject the gene of galactosidase into a female possessing a mutation that prevented it from producing this enzyme; then to find out whether the injected gene functioned in the female; whether the enzyme was synthesized; after how long a time; under what conditions.

The experiments were conducted by a biochemist from the University of California, Arthur Pardee, who had come to spend a sabbatical in Monod's department. Pink and blond, baby-faced, with a shy gaze behind his glasses, Arthur had an original mind that often diverged from the beaten track. His air of being a good little boy hid a remarkable experimenter: precise, quick, ingenious. Also more than a little finicky, for he insisted on using the same material under the same conditions as he had at Berkeley. Hence, there were a few difficulties in the first experiments we did together. Every minute, we each had to take a sample from a culture oscillating gently in a water bath. We turned, one behind the other, around the water bath. A sort of race, a frantic pursuit, with the least misstep letting loose an avalanche of errors. He loved the large flasks in which it was not easy to take samples. I measured the time of arrival of the gene in the females. He determined the quantities of enzymes they synthesized. A reaction that produced a yellow color; the more intense, the more enzyme there was. First experiment, first surprise. A big one! The gene functioned immediately, with no delay. As soon as it entered, a synthesis of the enzyme was observed at the maximum rate, at full speed. A puzzling result which did not fit with generally received ideas for explaining the synthesis of proteins. In the evening, in Monod's office, we pored over the results from every viewpoint, looking for a possible error, an overlooked blunder.

But this was merely the hors d'oeuvre. For the real problem that all three of us wanted to analyze, one that had fascinated

Monod for the past fifteen years, was the regulation of the lactose system. Most of the proteins in the colon bacillus were produced under all circumstances, whatever the conditions in the environment. This synthesis was called "constitutive." Galactosidase and permease, the proteins of the lactose system, were, however, formed only if the medium contained lactose or some analogue of this sugar. The synthesis of these proteins was said to be "inducible" by the lactose. Using inducible bacteria, however, Jacques had isolated "constitutive" mutants able to synthesize galactosidase and permease even in the absence of lactose. Finally, there was yet another type of regulation found by Monod's group and by other laboratories. It concerned the manufacture of essential constituents of bacteria, like the amino acids. Each of these compounds is manufactured in a series of reactions that, step by step, chisel the molecule to give it its final structure. In these chain reactions, the addition to the medium of the end product, the amino acid for example, stops the synthesis of the enzymes of the chain: a mechanism baptized "repression." For a teleologist, all these processes would have a precise significance: bacteria produce certain proteins only when they need to.

Studying the regulation of the lactose system required a long preparation. In itself, each experiment lasted barely an hour. But it took weeks to set one up: to obtain the series of bacterial strains we needed, with the desired genetic characteristics. Imperturbable and effective, Arthur Pardee saw to the details. All three of us spent several hours each day in Monod's office thinking up hypotheses, possible regulatory mechanisms, and inferring from them the results we could expect from projected experiments. What mattered most to Monod was to unify the different types of regulation. He had the art of generalizing from one particular situation, from an observation. To which was added an unshakable confidence in the logic of nature, in its unity. He had, thus, to find a single hypothesis to explain the two seemingly contradictory systems of regulation: induction and repression. Which could be obtained at the cost of a little intellectual gymnastics. Either by admitting that repression expressed the inhibition of a still unknown induction, a hypothesis called "generalized induction." Or, on the other hand, by considering that induction represented the blocking of a still invisible repression, a hypothesis called "generalized repression." It was the former hypothesis, generalized induction, that Monod favored.

Others, however, inclined toward generalized repression. That was the theme of a seminar that had just been given at the Pasteur Institute by the Hungarian-born physicist Leo Szilard. One of the fathers of the atomic bomb, Szilard had not endorsed its use against Japan. After Hiroshima and Nagasaki, he had given up physics to work in biology. Short, round, ruddy-featured, eyes darting behind his glasses, Leo Szilard was a truculent character crackling with insight and cleverness. Interested in everything, he had paradoxical solutions to every problem. At our first encounter, at a colloquium in the United States, he led me over to a corner to ask me about my work. At each response, he cut in to reshape my answers to suit his style, to force me to speak his language, to use his words, his expressions. He carefully noted each answer in a notebook. At the end, he said, "Sign there!" Two years later, during another encounter, he asked, "Is what you told me a few months ago still true?" And he noted: "Still true!" He traveled a great deal, going from laboratory to laboratory. A sort of fat bumblebee spreading ideas here and there like pollen. At the moment he was interested in generalized repression, whose merits he praised in his seminar. Against which Monod upheld the virtues of generalized induction. An inconclusive discussion, owing to the lack of decisive arguments.

It was experimentation that decided the matter. It took all winter. I have forgotten those days, with their incidents and their ups and downs, their measure of hope and difficulty. Except one. One fine afternoon, I had come down from the attic to the ground floor. In the laboratory I found Arthur and Jacques among dozens of test tubes containing assays of galactosidase, the yellow-tinted liquids. Tubes of intense yellow or clear yellow or pale yellow, which they were putting in order. And gradually the tendency emerged, became distinct. As in the tallying of votes on election night. As things clarified, the excitement grew. There is in research a unique moment: when one suddenly sees that an experiment is going to overturn the landscape. It is the moment when the facts combine to indicate a new and unforeseen direction. When the change taking place is due more to a feeling, to a premonition, than to the chilly logic of facts. Where the dream of novelty suddenly takes on consistency without being fully assured of becoming reality.

The problem was to specify the genetic relation between a synthesis of galactosidase called "inducible," which requires the presence of lactose, and a synthesis called "constitutive," which takes place even in the absence of this sugar. The day's experiment had consisted in mating an "inducible" male with a "constitutive" female whose galactosidase had been rendered inactive through mutation. Without an inducer, none of these strains could produce an active enzyme. As soon as the gene was transferred into the female, however, a constitutive synthesis of the enzyme was observed: a synthesis that ceased after half an hour and became inducible. All this seriously upset our way of thinking. First, because contrary to what had been believed up to that point, it was not the gene of galactosidase, but another closely linked, but distinct gene that ensured the regulation of the system. Next, because in a cell hosting both the inducible and constitutive forms of this gene at the same time, it was the inducible characteristic that predominated. The basic regulation was thus not induction but inhibition. It was the hypothesis of repression that won out. The "inducible" gene ensured the formation of a cytoplasmic product, immediately baptized "repressor," which inhibited the synthesis of galactosidase and of permease. The role of the inducer, lactose, was to inhibit the inhibitor, to repress the repressor.

These experiments, called PA JA MA (for Pardee, Jacob, Monod), marked a turning point in our way of representing the mechanisms of protein synthesis as well as those of regulation. Up to then, only "instructionist" theories had been proposed to explain the induced synthesis of galactosidase. The inducer, lactose, was supposed to "instruct" the protein: to tell it what form to take; to shape it according to certain folds that would confer on it its galactosidase activity. From now on, it would be necessary, on the contrary, to appeal to "selectionist" theories. The lactose acted like a signal for choosing the syntheses linked to its own metabolism, to activate the processes leading to the production of galactosidase and permease. Another victory of Darwin over Lamarck!

J. had phoned me to announce the next meeting of the Compagnons de la Libération. Important business. Come at any cost. Impossible, however, to give details on the phone. Someone might be listening in. Definitely, the people from the Resistance and the Free French would never change. Still the same taste for mystery. For secret services. Open secret. France was entangled in the Algerian mess. The Republic was threatened. The state was paralyzed. The government was in agony, a government headed by my old classmate at the Lycée Carnot, Félix Gaillard. Just about everywhere people were stirred up. A mixture of the Right, military men, veterans of Free France, supporters of French Algeria. Motions were proposed. Appeals to bring General de Gaulle back to power. This meeting of the Compagnons which J. announced was aiming, obviously, to rally the faithful and to appeal to the old leader of Free France. And, as a matter of fact, who else could rise to the challenge of the terrible Algerian drama? Who else could, once again, take in hand the nation's destiny? Still, good form required a minimum of legality, so as not to tarnish the old epic by a coup d'état, a return behind the paratroopers of Massu.

A lively meeting of some two hundred Compagnons, looking grave or agitated. Inflammatory speeches. Motions for the general's return to power. One or two timid objections. A stormy debate with a few bursts of fanaticism. Efforts to obtain unanimity. Finally the vote. Overwhelming majority. Only one vote against. After much hesitation, I abstained. A colonel from the paratroopers then came over to stare me up and down, challenging me.

Once admitted, once taught, science is cold. As cold as the techniques that derive from it. As cold as the texts explaining its content or the books reporting its history. Science in the works has two aspects: what could be called day science and night science. Day science employs reasoning that meshes like gears, and achieves results with the force of certainty. One admires its majestic arrangement as that of a da Vinci painting or a Bach fugue. One walks about in it as in a formal French garden. Conscious of its progress, proud of its past, sure of its future, day science advances in light and glory.

Night science, on the other hand, wanders blindly. It hesitates, stumbles, falls back, sweats, wakes with a start. Doubting everything, it feels its way, questions itself, constantly pulls itself together. It is a sort of workshop of the possible, where are elaborated what will become the building materials of science. Where hypotheses take the form of vague presentiments, of hazy sensations. Where phenomena are still mere solitary events with no link between them. Where the plans for experiments have barely taken form. Where thought proceeds along sinuous paths, tortuous streets, most often blind alleys. At the mercy of chance, the mind frets in a labyrinth, deluged with messages, in quest of a sign, of a wink, of an unforeseen connection. Like a prisoner in a cell, it paces about, looking for a way out, a glimmer of light. Ceaselessly, it goes from hope to disappointment, from exaltation to melancholy. It is impossible to predict whether night science will ever pass to the day condition. Whether the prisoner will emerge from the dark. When that happens, it happens fortuitously, like a freak. By surprise, like spontaneous generation. No matter where, no matter when, like thunder. What guides the mind, then, is not logic. It is instinct, intuition. It is the need to see clearly. It is the stubborn desire to live. In the endless interior dialogue, among the innumerable suppositions, connections, combinations, associations that constantly pass through the mind, a beam of light sometimes rends the obscurity. Suddenly the landscape shines with blinding light, terrifying, stronger than a thousand suns. After the initial shock begins a hard struggle with habits of thought. A conflict with the universe of concepts that governs our reasoning. One is not yet authorized to say whether the new hypothesis will get beyond its initial form of rough sketch and become refined, perfected. Whether it will withstand the test of logic. Whether it will gain admission to day science.

Late July 1958. A Sunday in Paris. The children have gone off on vacation. Lise and I have stayed home. She is at the piano in the next room, practicing a sonata. For my part, I am trying to get started on a lecture that I must give in New York. I have accepted too many engagements for the summer: a congress of microbiology in Stockholm, a congress of genetics in Montreal, a Harvey Lecture in New York. An honorific lecture that I want to get just right. My chosen theme: Genetic Control of Viral Functions. My heart is not in it. A day with no taste for work. With no desire to write this lecture. I go round in my study, chewing over vague hypotheses, possible experiments. At the end of the afternoon, fed up, weary, we decide to go to the movies. A film of no interest. Slumped in my seat, I dimly perceive in myself associations that continue to form, ideas for proceeding. An indistinct hullabaloo, whose unfolding I make no attempt to grasp. Shadows move on the screen. I close my eyes to heed the extraordinary things going on within. I am invaded by a sudden excitement mingled with a vague pleasure. It isolates me from the theater, from my neighbors whose eyes are riveted to the screen. And suddenly a flash. The astonishment of the obvious. How could I not have thought of it sooner? Both experiments—that of conjugation done with Elie on the phage, erotic induction; and that done with Pardee and Monod on the lactose system, the PA

JA MA—are the same! Same situation. Same result. Same conclusion. In both cases, a gene governs the formation of a cytoplasmic product, of a repressor blocking the expression of other genes and so preventing either the synthesis of the galactosidase or the multiplication of the virus. In both cases, one induces by inactivating the repressor, either by lactose or by ultraviolet rays. The very mechanism that must be the basis of the regulation. But there is more. With the phage it is not simply two or three proteins whose synthesis is blocked, but at least fifty. And also the reproduction of the chromosome of the phage, of a whole fragment of DNA. Where can the repressor act to stop everything all at once? The only simple answer, the only one that does not involve a cascade of complicated hypotheses, is: on the DNA itself! In one way or another, the repressor must act on the DNA of the prophage to neutralize it, to prevent the activity of all its genes. And by way of symmetry, the repressor of the lactose system must act on the DNA containing the genes of the galactosidase and of the permease.

These hypotheses, still rough, still vaguely outlined, poorly formulated, stir within me. Barely have they emerged than I feel invaded by an intense joy, a savage pleasure. A sense of strength as well, of power. As if I had climbed a mountain, attained a summit from which I saw in the distance a vast panorama. I no longer feel mediocre or even mortal. I need air. I need to walk. I stir in my seat. Lise looks at me, intrigued: "You've had enough? You want to leave?" We find ourselves on the boulevard Montparnasse. Then I am seized with a sort of intoxication. An immense need to talk. To sharpen my ideas by relating what is exciting me. I say to Lise: "I think I've just thought up something important." She looks at me, eagerly: "Tell!" Once back home, I try to explain to her as simply as I can the facts of the problem. The two situations: their similarity; even identity. My conclusions. The regulation on DNA itself. The importance of this hypothesis. She listens attentively, but the conclusion disappoints her: "You've already told me that. It's been known for a long time, hasn't it?" I venture a slightly sad smile before going back in my shell. Whom to talk to? Monod is off on vacation. Lwoff, too. I do not know where Wollman is. I am alone with my thwarted dream.

A new kind of exercise awaits at New York University: the lunch seminar. For an hour I had to tell my story about bacterial conjugation to some fifty grimacing faces, jaws working away. From time to time in the middle of a mouthful, someone in the audience would spew out a question. After the talk I was allowed a sandwich while I answered questions. At the end, a tall, husky student twenty or twenty-two years old sat next to me, chewing gum: "I was very interested in your genetics experiments. But I would like to know what you think of the way the French are treating Algeria. The Algerians are in their own country, aren't they? So why kill them and burn their villages? And wouldn't it get even worse with General de Gaulle?" Cold fury seized me. For while I did not approve of France's actions in Algeria, I approved even less of being lectured to by this boor. I thought the atrocities committed by the French troops were indefensible and Algeria's independence inevitable. But, faced by the arrogance of this gum chewer, I felt obliged to maintain the opposite, to justify the mistakes and horrors we were committing there. Which pained me!

When I entered Monod's office that afternoon in September 1958, I was both worn out and excited. I had returned from New York the night before without getting a wink of sleep. After lunch at home with Lise and the quartet, I hurried to the lab to see Monod. Since I had not seen him all summer, I had had no chance to explain the ideas that had suddenly come to me some weeks earlier at the movies. In fact, I had not yet tried out my hypothesis on anyone. But all summer I had mulled it over, ruminated on it, refined it at every opportunity, during plane flights, during nights in one hotel or another. The more I thought about these ideas, the more they intoxicated me. I had even reached the point where they seemed hard to refute, for I found in them both a coherent explanation and a suggestion of further avenues to explore. I still had to get Monod to accept this business, for he held all the keys to the lactose system: the spoken and the unspoken, the breviary and the folklore. Everything depended on him. Hence my excitement. After a few words of welcome and friendly banter, I began to tell my story. But my eyes were stinging. I was dead tired. I spoke with the volubility that often comes from a mixture of fatigue and jangled nerves. And Jacques was hardly listening to me. At first he smiled faintly. Then he burst into laughter, one of his enormous laughs that filled the whole building, allowing anyone at a distance to know precisely where he was. Childish! My model was altogether childish! Off the top of his head, without even thinking about it, he could give me on the spot a series of arguments any one of which was enough to destroy the model. Immediately my fever broke. My fatigue won out over my excitement. I decided to put off to the next day any further attempt at discussion. I would go home and get some sleep.

Obviously, in the state I was then in, I could only fail to reach my goal. My speech had nothing seductive about it. While, for his part, Monod had little desire to be seduced. He did not, in fact, like to change hypotheses. And when he did change them, he liked the change to come from himself, not someone else. As is often true in such a rich and ardent nature, several distinct and even opposed characters coexisted in Jacques Monod. Two of them as far as the man of science was concerned. The first—call him Jacques—was a man full of charm, warmth, and generosity. A man who, loving people as much as ideas, was always available to his friends, always ready to listen and to help. A man of exceptional gifts who was interested in everything. A rigorous and penetrating mind, not hesitating to criticize himself, always asking good questions. The second persona—call him Monod—was dogmatic, very sure of himself, and domineering. A man who wanted to attract attention, lusting after praise and admiration, trying to shine in every field. A man given to pronouncing definitive judgments on everything and everyone, who loved to show his colleagues the true meaning of their own work, but dismissed as stupid even the most tentative objection. Without the slightest hesitation, Jacques would drop his affairs to go hundreds of miles to help a friend in trouble. With a few phrases, Monod could make a friend into an enemy. Working with Jacques was a pleasure. Debating Monod could be trying. Happily, at the time it was Jacques who most often prevailed.

On my return from New York, I had been received by Monod. But the next day, when after twelve hours' sleep I re-entered his office, I found Jacques. A receptive Jacques, who listened to me with attention and curiosity. Seeing him interested increased my energy. I launched into my story, to defend my arguments. To make analogies between lysogeny and the lactose system, between erotic induction and the PA JA MA experiment. Analogies so close that one could not reject the hypothesis of a common mechanism. Each of these systems had its advantages and its drawbacks. With phage, genetic analysis was not too difficult, but proteins were not well understood. With the lactose system, on the other hand, it was not hard to measure the enzymatic activities of proteins, but the genetics was imprecise. Hence, they were experimentally complementary. And once a common mechanism was postulated, lysogeny imposed certain constraints on the possible models. Because, notably, of the fifty proteins of the phage whose synthesis was completely blocked by a single repressor. As if this repressor acted on a single element required for all these syntheses. As if it closed a single lock to bolt in a single action the activity of the whole chromosome, of all the phage genes. Which, for me, had two consequences. First, the idea of genetic units that were more extensive than the gene, "units of activity" containing several genes submitted to a common expression, to a common regulation that probably occurred on the DNA itself. Next the idea of a regulation operating, not progressively, but discontinuously, like a switch, by an all-or-none, stop-or-go mechanism.

All this time, Jacques had said nothing. Just one or two questions to clarify certain points. Clearly, he was interested. Without, however, appearing disposed to swallow it whole. After a long silence, he began to go over it point by point, detail by detail. At first, this story of regulation functioning directly on the DNA did not please him. Perhaps, he added, simply because of his preconceived ideas. For a mind trained in classical genetics, a gene or a chromosome represented an intangible structure: a sacrosanct object that could not actually be manipulated without attacking life itself. But after all, in bacteria, it had just been possible to prove the contrary with the prophage and the episomes; to show that entire fragments of a chromosome could be added or taken away without grave consequences. Why not then a regulation on the gene itself? An idea not to forget!

Jacques was in his element on the theme of the switch, the stop-or-go regulation. On kinetics, the speeds of synthesis, their variations with the quality of the inducer and its concentration, he was unbeatable. He had arguments for every situation, answers to every question. And there, he was adamant: there was no chance of reconciling these results with an all-or-none mechanism. Nevertheless, I was unconvinced. Perhaps out of ignorance, I liked this hypothesis. First because of its simplicity. Also for a crazier reason. Some weeks earlier, I had observed my son Pierre playing with a model electric train. This train had no rheostat. Nevertheless, Pierre could vary the speed of the train by manipulating the switch, making it oscillate faster or slower between start and stop. Then why not a similar mechanism in the synthesis of proteins? Jacques did not even want to discuss such an argument. For him, it was a joke. Moreover, a bad joke.

Nevertheless, I had a sense the "fish was biting." That my interlocutor was caught in the game. As the discussion progressed, Jacques became animated. He got up. Paced back and forth. Halted in front of the blackboard. Drew a diagram. Stood meditating for a moment. Took his chin between two fingers. Went back and sat down. Kept silent for a while. Began thinking out loud. Without any doubt, the machine was running. An impressive machine, constructed with rigor and logic. With precision in its reasoning. Which very quickly led Jacques to an important remark: to the need for an inference that had utterly escaped me. If the two genes, that of galactosidase and that of permease, constituted the same unit of activity, submitted to the same regulation, the synthesis of the two proteins had to be coordinated. Whatever were the conditions, the situations, the two activities could appear only in the same ratio. Which had almost always proved to be the case, added Jacques firmly. Almost always, but not quite always! In two cases, at least, there was no coordination. In particular, with a certain analogue of lactose that induced the synthesis of galactosidase, but not of permease. This kind of objection was exactly what I feared, but without particularly caring about it. More important now was Jacques's growing interest in this dialogue. Obviously he was reacting. He was hooked. He looked for arguments as much on one side as on the other. I had a sense of having largely won. Of seeing my hypotheses take on consistency, substance. Go gradually from night science to day science. It was, however, growing late. The discussion had exhausted me. Unlike Jacques, who was as fresh and alert as in the morning. It was time to stop. To find the other members of the group gathered around a bottle of Scotch.

That day marked a turning point in my scientific life. Some months before, my long collaboration with Elie Wollman had been completed. His thesis defended, Elie had left for Berkeley to spend a year in Gunther Stent's lab. While I had seen Elie in New York that summer and had tried to talk to him about the hypotheses that were exciting me, he paid not the least attention. A little sad, I had accompanied him and his wife, Odile, to the train they were taking for California. That day in Jacques's office began a new era. An era of activity, of rare exaltation. The most intense period of my scientific life. From having first been reserved about the model, Jacques came to take a close interest in it. His criticisms grew more and more constructive. He dropped some of his objections. Others persisted: notably the absence of coordination in the induction by a certain analogue of lactose, which remained a skeleton in the closet. But we agreed to forget it for a while.

Each day, we had long discussions. Each of us tried out new ideas on the other. Jacques concentrated on the lactose system. For my part, fascinated with their similarity, I endeavored to consider the two systems together. Detecting a new mutant in one system led directly to the prediction and isolation of the corresponding mutant in the other system. During one of our sessions, Jacques made an important observation: if there was a switch activated by the repressor, it had to be specific, so as to be modifiable by mutation. Immediately we sought to define the properties of such a mutation; to draw on the blackboard our representation of the regulatory circuit. We saw this circuit as made up of two genes: transmitter and receiver of a cytoplasmic signal, the repressor. In the absence of the inducer, this circuit blocked the synthesis of galactosidase. Every mutation inactivating one of the genes thus had to result in a constitutive synthesis, much as a transmitter on the ground sends signals to a bomber: "Do not drop the bombs. Do not drop the bombs." If the transmitter or the receiver is broken, the plane drops its bombs. But let there be two transmitters with two bombers, and the situation changes. The destruction of a single transmitter has no effect, for the other one will continue to emit. The destruction of a receiver, however, will result in dropping the bombs, but only by the bomber whose receiver is broken. All situations that gave mutations whose properties were very precise and well defined in genetic terms. The mutation of the transmitter was called "recessive"; that of the receiver "cis-dominant," that is, on the same chromosome. During this development, our voices had risen. The excitement grew. We were drawing more and more feverishly on the blackboard. All kinds of diagrams, with arrows going in every direction. And suddenly I realized that several years ago Elie Wollman and I had isolated and studied a phage mutant that showed exactly the properties expected of the receiver of the repressor. How stupid I was not to have thought of it sooner! At that point, our joint confidence in the model grew by a factor of a thousand.

Jacques and I collaborated ever more intensely each day. On the ground floor, in the laboratory adjacent to his office, inducers were added, activities assayed, syntheses measured. In the attic, in the room I had in André Lwoff's lab, mutations were isolated, characterized, localized. I spent a good part of my day going up and down the stairs carrying test tubes or plates of culture. Almost invariably, each day ended in Jacques's office where we had an interminable discussion of the model, looking for variations or for opportunities to demolish it. We divided up the experimentation, each of us to explore some particular aspect of the model. For Jacques and his team, the first job was to make precise the coordination of the synthesis. As foreseen, one always found the same ratio between the two activities, galactosidase and permease, in every condition. In all except one, always while using as an inducer the same analogue of lactose. Once again, it was necessary to keep the door shut on the skeleton in that closet. As for me, my job was to isolate the "constitutive" cis-dominant mutant of the lactose system. The mutant in which the modified switch had become unresponsive to the repressor. A difficult trick, for it required using a colon bacillus with two copies of the chromosome and containing a double example of the genes of the lactose system. Which did not exist, for the colon bacillus has only one copy of its chromosome. So it was necessary to invent it, using the infinite resources of microbial genetics. Certain genetic units, which Elie and I had called "episomes," behaved like little chromosomes. One could at will add them to a bacterium or draw them out. In a few months, we learned, with Martine Tallec and Danièle Charpentier, to insert selected genes in these units: notably the genes of the lactose system. Thus there were two sets per bacterium: one on the chromosome, the other on the episome. Nothing easier, then, than to isolate dominant constitutive mutants. To my great joy, these mutations activated only the gene of the galactosidase placed in "cis" on the same chromosome, just the property required of the switch. Again to my great joy, they simultaneously activated the two adjacent genes, that of galacto-sidase and that of permease. The "units of activity" were not a daydream. The lactose system formed one of these units.

The exaltation that had followed the first fleeting vision of the model during the summer had soon been mingled with a vague malaise: a sense of having discovered, even of having stolen, a secret both untransmittable and sacred. A secret that violated certain principles that had been inculcated in me. Perhaps because of the impossibility of my discussing these ideas for the whole summer, of having to keep them to myself, to mull them over on interminable trips, it had seemed to me, at first, that I had no right either to go into them more deeply or to speak of them lest they let loose on me the thunderbolts of some mysterious power. After the first long discussion with Jacques, these impressions had given way to wonder. Wonder, first, at feeling myself to be participating actively, contributing, despite my ignorance and delay, in what seemed to be one of the great adventures of the century. Wonder also at seeing gradually take form what had been only a dream, a sort of mental game, a pure product of the imagination springing up in the darkness of a movie theater. Now to become a subject for discussion by the most serious scientists. To guide experimentation. To be confirmed on some points, refuted on others. To be modified bit by bit. In short, to acquire a form, a consistency, even a material existence, like an architect's vision that materializes in the construction of a palace. I felt reality flowing by in a dream. Evidently, the world of science, like that of art or religion, was a world created by the human imagination, but within very strict constraints imposed both by nature and by the human brain. As if this science endeavored not to photograph nature, but to paint it. To decompose it in order to refashion it by every means at its disposal. To obtain from it a representation of a logical truth, especially of a possible and communicable truth for whoever was willing to go to the trouble of looking at it.

Wonder again at having, with this model of gene regulation, penetrated one of the mysteries of life. Of having reached the very essence of things. Of having gained access to a primordial mechanism. A mechanism fundamental to all living beings from their very beginnings, and that would persist as long as they exist. And with this idea that the essence of things, both permanent and hidden, was suddenly unveiled, I felt emancipated from the laws of time. More than ever, research seemed to be identified with human nature. To express its appetite, its desire to live. It was by far the best means found by man to face the chaos of the universe. To triumph over death!

Wonder, finally, at being engaged in a long duet with Jacques Monod. With a man who had long appeared a sort of hero, an unattainable model, a personage standing on the heights that, for me, would remain forever inaccessible. From my student years in medicine, I had retained a certain stiffness, a respect for hierarchy that made me call André Lwoff "Monsieur" and raised a barrier around Jacques Monod. A barrier that fell when our daily collaboration became as close, as intertwined, as it was possible to be. Very quickly Jacques had taken over the model. He had made it his, modifying it at some points to conform with certain facts. Putting all his rigor, all his intelligence into defining different aspects of it, in sorting out all the consequences. Between us we wove an intellectual relationship of exceptional intensity, even intimacy. Each day we spent several hours together, morning and evening, computing the results of experiments, analyzing them, criticizing them, drawing inferences, rectifying hypotheses, concocting new experiments. Singing a cantata, whistling a quartet. Also joking. For all this went on in an atmosphere both feverish and gay where each teased the other at every turn: about his origin, medical student or zoologist; on his "general from the war," de Gaulle or de Lattre de Tassigny; on his culinary or political or literary tastes. All this among huge bursts of laughter that shook the new office Jacques had moved into after the death of Gabriel Bertrand: a large corner room in the chemistry building, with large windows overlooking the garden, with light-colored furniture and woodwork, the couch and the armchairs upholstered in brick-red. And, at the end of the room, the large dark-green blackboard where we furiously sketched formulas and diagrams. Very quickly we had reached a rare degree of complicity. Our discussions moved at top speed, in bursts of brief retorts, like a ping-pong match. Scarcely had one begun a sentence than the other answered without waiting for the end. Which often discouraged and confounded people who joined in our discussions but, less knowledgeable about our problems, our mutants, our codes, had trouble keeping up with us. The harmony between us was so close, our repartee so quick, our train of ideas so coordinated, that it was often difficult to say which was the first to come up with a hypothesis or to propose an experiment. As the music of a choir may be so coherent, so integrated, as to make individual voices indistinguishable. In the duet, each showed himself as he was, with his pride, his ambition. But without makeup, without theatrics. For close to five years, our friendship unfolded like an epic poem.

17 June 1959. The reception hall of the Elysée Palace. A sort of great conservatory, like the Saint-Lazare train station, all crimson and gilt. Back in business, having become president of the Republic, General de Gaulle receives the Compagnons de la Libération on the anniversary of 1940. A short welcoming speech. The general has become heavy-set. The Gothic cathedral's ribs have lost their curve and the arches their delicacy. But his chin remains high, his gaze fixed on the horizon. And his voice has lost nothing of its fire or its power of seduction and domination. The return to power has taken place with due ceremony. The general has seized the reins with as firm a hand as ever. He has even understood the importance of research and has put into effect the recommendations of the colloquium in Caen, until then a dead letter.

At the moment, he is receiving the Old Guard. Some three hundred Compagnons come from all over France. Many of my old comrades from Free France. Several politicians from the Resistance. A buffet. Champagne. Petits-fours. The general crosses the room, going from group to group. A word for each. A longer word with the parents of Compagnons who have died. Suddenly I am before him. I present myself. He extends his hand.

"Ah, Jacob. Pleased to see you again." Pause. "What are you doing these days?"

"Scientific research, sir," I stammer.

"Ah, very interesting. In what field?"

"Biology, sir."

"Ah, very interesting. What kind?"

"Genetics, sir."

"Ah, very interesting. And where do you work?"

"At the Pasteur Institute, sir."

"Ah, very interesting. You have everything you need?"

After a hesitation: "No, sir."

"*Au revoir!*"

There was nothing to be done. We were not succeeding. In vain did we try to check through the experiment, to modify it, to change a detail here and there. It was now three weeks since Sydney Brenner and I had arrived at the California Institute of Technology. We had come for the sole purpose of carrying out this experiment with Matt Meselson. An experiment that we had no doubt was going to change the world. But the gods were still against us. Nothing worked. Our fine confidence at the start had evaporated. Disheartened, Meselson had departed—to get married! Sydney and I talked about going back to Europe. In a burst of compassion, a biologist by the name of Hildegaard had taken us under her wing and, to give us a change of scene, driven us to a nearby beach. There we were, collapsed on the sand, stranded in the sunlight like beached whales. My head felt empty. Frowning, knitting his heavy eyebrows, with a nasty look, Sydney gazed at the horizon without saying a word.

Never yet had I seen Sydney Brenner in such a state. Never seen him silent. On the contrary, he was an indefatigable talker at every opportunity. A tireless storyteller, able to discourse for days and nights on end. Interminable monologues on every conceivable subject. Science, politics, philosophy, literature, anything that cropped up. With stories he made up as he went along. Generously laced with jokes. With nasty cracks, too, at the expense of just about everyone. An excellent actor, he could render a speech in Hungarian, a lecture in Japanese. Mimic Stalin or Franco. Even himself. He went without a break from one register to another. A sort of fireworks whose effects he gauged from the expressions of the people around him. I had first met him at a colloquium in the United States. Short, broad of back, this character hardly went unnoticed. His squarish head, his eyes blue beneath blond brows, enormous, hirsute, shaggy, he resembled certain Dutch portraits. A Franz Hals. But behind his slightly sarcastic, even satanic visage, his smile revealed a child's face. And thought, which often causes a frown, gave his features the serenity of sleep. Born in South Africa, he had settled in Cambridge University's Laboratory of Molecular Biology, already home to Fred Sanger, Max Perutz, John Kendrew, and Francis Crick. A beautiful string of prima donnas into which Sydney fitted perfectly.

This trip to Pasadena, this fiasco that had led us to a Pacific beach, was the sequel to a story begun two years earlier in our laboratory. The findings of the PA JA MA experiment and the study of regulation were difficult to reconcile with the representation we then had of protein synthesis. We knew that protein synthesis took place in the cytoplasm, on tiny granules called "ribosomes." It was considered that for each gene there were corresponding ribosomes specifically charged with producing the corresponding protein. But these ribosomes, made up of proteins and RNA, were very stable structures, lasting for several generations. A scheme that accorded neither with the synthesis of galactosidase immediately upon entry of the gene. Nor with the existence of units of activity recently baptized "operons," that contained several genes. Nor with a regulation functioning directly on the DNA through the intermediary of a switch, now called an "operator." Hence, the perplexity prevailing in the

Pasteur group; the interminable exchange of letters with Arthur Pardee in Berkeley; the endless discussions in Paris with Jacques and François Gros, our specialist in RNA. Hence, too, the wavering between the two possible hypotheses: either direct synthesis of the protein on DNA itself, with no intermediary; or production of an unstable intermediary, probably an RNA with rapid renewal. But the former hypotheses seemed highly improbable and the latter without a chemical basis, without any trace of a molecule that could substantiate it. Even so, this was the hypothesis we preferred.

I had related this story, and our doubts, at a small colloquium on microbial genetics organized in Copenhagen by Ole Maaloe in September 1959. A small group attended, including notably Jim Watson, Francis Crick, Seymour Benzer, Sydney Brenner, Jacques, and even the physicist Niels Bohr. Courteous as ever, Jim Watson spent most of the sessions ostentatiously reading a newspaper. So, when it came time for him to speak, everyone took from his pocket a newspaper and began to read it. I had been assigned to lay forth the views of the Pasteur group: the PA JA MA experiment; the operon; the operator. I stressed the difficulty of reconciling these results with the current model of protein synthesis. I invoked the two other hypotheses by stressing the need for an unstable intermediary which I called X. No one reacted. No one batted an eyelash. No one asked a question. Jim continued to read his newspaper.

A new opportunity to discuss protein synthesis arose around Easter 1960 in Cambridge, in Sydney's small apartment in King's College, of which he was a Fellow. With only Francis Crick, Leslie Orgel, Ole Maaloe, and an American researcher. Francis and Sydney wanted to discuss in detail our experiments, which had been published only in the form of short notes in French. They made me take a veritable examination! With questions, criticisms, comments. A pack of hounds racing around me, nipping at my heels. I was, however, in a position of strength. If it hadn't been for their rapid-fire English, I would have felt quite at ease. All the more so as I had new results to report: an experiment long prepared in Paris and recently completed in Berkeley by Arthur Pardee and his student Monica Riley. They had succeeded in charging the DNA of male bacteria with radioactive phosphorus; in making them transfer to females the gene of galactosidase; in letting it synthesize the enzyme for some minutes; and then in destroying the gene through the disintegration of the radioactive phosphorus. The result was clear: once the gene was destroyed, all synthesis stopped. No gene, no enzyme. Which excluded any possibility of a stable intermediary.

At this precise point, Francis and Sydney leaped to their feet. Began to gesticulate. To argue at top speed in great agitation. A red-faced Francis. A Sydney with bristling eyebrows. The two talked at once, all but shouting. Each trying to anticipate the other. To explain to the other what had suddenly come to mind. All this at a clip that left my English far behind. For some minutes, it was impossible to follow them, just as it would have been impossible for them to follow a discussion in French between Jacques and me. What had set off Francis and Sydney was, once again, a connection between the lactose system and phage. After infecting the colon bacillus, certain highly virulent phages blocked the synthesis of new ribosomes. As had been shown by two American researchers, Elliot Volkin and Lazarus Astrachan, the only RNA then synthesized had two remarkable properties: on the one hand, unlike ribosomal RNA, it had the same base composition as DNA; on the other hand, it renewed itself very quickly. Exactly the properties required for what we called X, the unstable intermediary we had postulated for galactosidase. Why, in Paris, when we were looking for a support material for X, had we not thought of this phage RNA? Why had I not thought of it? Ignorance? Stupidity? Oversight? Misreading of the literature? Failure of judgment? A little of all these, no doubt. A mixture that, as in a detective novel, had made us fail to spot the murderer, the molecule responsible. In the last analysis, however, what mattered was that X, the unstable intermediary, was materializing. That what until then had been an abstraction was becoming attached to a molecular species. Once again, a functional requirement had led to the search for a substrate. Once again, a creature of pure reason came to life.

Again, it had to be shown that all this was not a dream; that this RNA of the phage was indeed the unstable intermediary functioning in the synthesis of proteins: the issue that we and Sydney immediately decided to take up. That evening, the Cricks were giving a party. A very British evening with the cream of Cambridge, an abundance of pretty girls, various kinds of drink, and pop music. Sydney and I, however, were much too busy and excited to take an active part in the festivities. We isolated ourselves in a corner with beer and sandwiches to lay our plans. In the afternoon, we had discovered that we had both been invited to spend the month of June at the California Institute of Technology: Sydney by Matt Meselson, I by Max Delbrück. A unique opportunity to work together to demonstrate the nature and role of X. Especially as the experiments to carry out this end required a technique recently developed by Meselson. Since the morning session at King's College, we had been forming a new representation of protein synthesis. The ribosomes had lost all specificity. They had become simple machines for assembling amino acids to form proteins of any kind, like tape recorders that can play any kind of music depending on the magnetic tape inserted in them. In protein synthesis, it was X, the unstable RNA copied on a gene, that had to play the role of the magnetic tape, associating with the ribosomes to dictate to them a particular sequence of amino acids corresponding to a particular protein. Hence, the following experiment: to show that the unstable RNA, synthesized *after* infection of a colon bacillus by the virulent phage, associated with pre-existing ribosomes, synthesized *before* infection, to produce the proteins of the phage. Simpler to say than to do!

It was difficult to isolate ourselves at such a brilliant, lively gathering, with all the people crowding around us, talking, shouting, laughing, singing, dancing. Nevertheless, squeezed up next to a little table as though on a desert island, we went on, in the rhythm of our own excitement, discussing our new model and the preparations for experiments at Caltech. A euphoric Syd-

ney covered entire pages with calculations and diagrams. Sometimes Francis would stick his head in for a moment to explain
what we had to do. From time to time, one of us would go off for
drinks and sandwiches. Then our duet took off again. At every
moment, we found some new argument to support the theory;
rarely, one against it. In the light of our model a whole series of
different phenomena, of observations from many fields, of
seemingly unrelated facts, fell into place, fitting together like
pieces of a jigsaw puzzle.

Arriving on the West Coast a few weeks later, I gave seminars
at several universities: Berkeley, Stanford, Caltech. My story was
generally well received. Everything was readily accepted: the PA
JA MA experiment and the synthesis of galactosidase; the analysis of the regulation; the operon; the operator. Everything but the
hypothesis of the magnetic tape coming to be associated with the
ribosomes and dictating to them a specific program of protein
synthesis. People bristled at this idea, rolled their eyes in horror.
With a little encouragement, my audience would have jeered and
left. Those who did not like the Pasteur group sneered. Those
who did like us strongly advised me to talk about something else.
As for Max Delbrück, he simply threw up his hands.

In Delbrück's department at Caltech, Sydney and I set up shop
in the lab of Jean Weiglé, who was then in Europe. A small,
rather somber room which made us feel we were in exile. Matt
Meselson, who was working in the chemistry department, came
in as a reinforcement. He was a slender, dark-haired young man
with a fine-featured, smiling face and serious gaze, and was given
to precise speech. He was haunted by the Cold War, by the need
to establish better relations with the Soviet Union. In his soft
voice, he could discourse for hours on strategy, tactics, nuclear
arms, the Rand Corporation, first strikes, reprisals, annihilation.
In biology, he had had a brilliant idea that had allowed him to
resolve a series of problems. He marked macromolecules by cultivating bacteria in heavy isotopes before putting them back in a
normal environment. Using ultracentrifugation, he could then
separate the marked molecules along gradients of density and

track their synthesis, their stability, their evolution. This was the technique that Sydney and I wanted to use in our experiment: to mark the ribosomes with heavy isotopes before the infection of bacteria by the phage. To mark the RNA with radioactive isotopes after infection. If we were right, if our hypothesis was correct, the radioactivity of the RNA had to be associated, in the gradients, with the band of "heavy" ribosomes. We had given ourselves a month to do this experiment. A month that neither of us could extend. One month to succeed!

This was the first time I had worked abroad or even in a laboratory other than in the Pasteur Institute. I felt awkward, hampered. I could find nothing: not the pipettes, not the chemicals, not even the distilled water. Worse than a beginner. Sydney was in his element. He was at home inside two hours. He had located everything, prepared the medium, seeded the bacteria, begun an experiment. We were to do very long, very arduous experiments. One part was carried out in Weiglé's lab; another in the basement where centrifuges and Geiger counters were located. There were interminable dead periods while the centrifuge was spinning and the density gradient was forming. After which, we had to recover the contents of the centrifuge tubes, drop by drop; then, for each drop, to measure the absorption of ultraviolet light and count the radioactivity. We settled in front of the Geiger counter as though it were a television set, and watched the figures appear. But nothing worked. We had tremendous technical problems: in cultivating the bacteria in the presence of heavy isotopes; in preparing a sufficient number of ribosomes; in preventing them from decomposing. The results had no coherence. Instead of arranging itself in a well-defined peak, the radioactivity was distributed randomly. In every different way except the one we wanted. From time to time, Max Delbrück would peer around the door, raise his eyebrows inquiringly, and in a sarcastic tone ask for news of X.

Full of energy and excitement, sure of the correctness of our hypothesis, we started our experiment over and over again. Modifying it slightly. Changing some technical detail. Unstoppable,

Sydney introduced some new procedure every day: to force the bacteria to accept heavy isotopes; to fix the ribosomes; to obtain better gradients. But without great success. The results showed a bit more coherence. The radioactivity dispersed a bit less. A few shadows of peaks showed up. But not where we were expecting them, not with the heavy ribosomes. Despite our best efforts, our discussions, our critical examinations of the results, we could not find a flaw in the design and execution of our experiments. We spent hours waiting for the centrifuge to stop. Counting samples. Still without the hoped-for result. Max continued to look in on us, his tone each day more ironic. Matt would come by to encourage us. But as time passed, his enthusiasm waned. And our time was running out. For, come what may, Sydney and I had decided to leave at month's end.

Nineteen-hundred sixty was a presidential election year in the United States. The parties were preparing for their conventions, the favorites being, on the one side, Kennedy, and on the other, Nixon. Which unleashed passions in American universities, which were largely committed to Kennedy. From afar, we followed campaign developments, on television or through what the Caltech researchers told us. These researchers were attentive hosts. Lunches. Dinners. Parties right and left. A musical evening in which four stars of the biology department made up a string quartet. Everyone took a distant interest in our experiments. Came to visit us rather as one goes to a zoo. Worried about our morale, our progress. Asked with commiseration for news of X, in whose existence not one of them, of course, believed. The days passed. Our stock of time dwindled. And despite our relentless efforts, the experimental results were not panning out. Our confidence crumbled. Worn out, Meselson abandoned us to rejoin his fiancée.

That is why, thanks to the solicitude of the biochemist Hildegaard, we found ourselves lying limply on a beach, vacantly gazing at the huge waves of the Pacific crashing onto the sand. Only a few days were left before the inevitable end. But should we keep on? What was the use? Better to cut our losses and return home. Curled in on himself, Sydney exhibited the closed mask of a bulldog. From time to time, one of us repeated the litany of the failed manipulations, trying to spot the flaw. A good woman, Hildegaard tried to tell us stories to lighten the atmosphere. But we were not listening. Suddenly, Sydney gives a shout. He leaps up, yelling, "The magnesium! It's the magnesium!" Immediately we get back in Hildegaard's car and race to the lab to run the experiment one last time. We then add a lot of magnesium. In my haste, I miss a tube with my pipette which then spills a huge quantity of radiophosphorus into Weiglé's *bain-marie*. A short time later we tiptoe down to the basement to conceal the contaminated *bain-marie* behind a Coca-Cola vending machine.

Sydney had been right. It was indeed the magnesium that gave the ribosomes their cohesion. But the usual quantities were insufficient in the density gradients used to separate heavy and light compounds. This time we added plenty of magnesium. The result was spectacular. Eyes glued to the Geiger counter, our throats tight, we tracked each successive figure as it came to take its place in exactly the order we had been expecting. And as the last sample was counted, a double shout of joy shook the basement at Caltech. Followed immediately by a wild double jig.

This was merely one experiment, performed in extremis. Lacking controls. Also lacking the proof that these bacterial ribosomes, where the phage unstable RNA had attached itself, did indeed produce the phage proteins. Long and difficult experiments that Sydney was to complete in Cambridge. But we now knew that we had won. That our conception explained the transfers of information in the synthesis of proteins. Sydney had resumed his interminable monologues and jokes, his explosive outbursts and spiteful cracks. Amazing Sydney. Always playing with ideas, things, words. A mind as agile as his hands. Scarcely was the experiment over than we gave a seminar at Caltech to demonstrate the existence of X and its role as magnetic tape. No one believed us. The next day we left, each to his own home. The bet had paid off. In the nick of time.

Jacques's office. Again, endless discussions at the blackboard. But this time about writing. About putting in order this mass of data gathered over three years. About giving it shape. Creating a story out of it: the official transcript of this research. A story with enough force, enough persuasiveness to convince the others. To get them to adopt our point of view and shed light on their own research.

A strange exercise. Science is above all a world of ideas in motion. To write an account of research is to immobilize these ideas; to freeze them, like describing a horse race with a snapshot. It is also to transform the very nature of the research; to formalize it. To substitute an orderly train of concepts and experiments for a jumble of disordered efforts; of attempts born of a desperate eagerness to see more clearly; and also of visions, dreams, unexpected connections; of simplifications often childish, random soundings in every direction, never knowing where one is going to end up. In short, writing a paper is to substitute order for the disorder and agitation that animate life in the laboratory. Nevertheless, as the work progresses, how can we not seek to acknowledge the roles of chance and of inspiration? But to get some work accepted and a new way of thinking adopted, it is necessary to purify the research of all affective or irrational dross. To get rid of any personal scent, any human smell. To proceed on the royal way that leads from babbling youth to blooming maturity. To replace the real order of events and discoveries by what appears as the logical order, the one that should have been followed if the conclusion were known from the start. There is a ritual in the manner of presenting scientific results. Rather as if one were writing the history of a war using only the official press releases of the general staff.

Jacques was a past master at this game. After long discussions of what content and form to give this article, we had each written our part. But my English was stiff and awkward. Jacques's English was clear and flamboyant. He wrote the whole final version. Our review depicted like a fresco the study of protein synthesis and its regulation. It fitted our work into a historical perspective.

It integrated the data gathered on the galactosidase, on biosynthetic pathways, on the phage. It included the efforts made with Sydney in Pasadena, and by François Gros and Walter Gilbert in Jim Watson's laboratory at Harvard, to give content to X, now renamed "messenger RNA." It showed for the first time how a gene functions; how it sends a continuous stream of information toward the cytoplasm, rather like a faucet whose flow can be regulated according to the requirements of the cell as a function of signals from the environment. It proposed a model to explain one of the oldest problems in biology: in organisms made up of millions, even billions of cells, every cell possesses a complete set of genes; how, then, is it that all the genes do not function in the same way in all tissues? That the nerve cells do not use the same genes as the muscle cells or the hepatic cells? In short, this article presented a new view of the genetic landscape.

On one point there was a difference between Jacques and me. A difference in personality, in our attitudes toward nature. It happened that, in the three systems studied in the laboratory—lactose, phage, and the biosynthetic pathway of tryptophane—regulation functioned in the same way. In the three cases, we had observed what we called a "negative" regulation, where the expression of the genes was inhibited by a repressor, as opposed to what might be called "positive" regulation, where this expression would require the presence of an activator. This unity among the three mechanisms gave Jacques great satisfaction. He had even decided that this unity had to extend to the entire living world, that everything had to function in the same way by exclusively negative systems of regulation. Such uniformity seemed to me improbable. To achieve performances as astounding as the production of a human being from an egg, it seemed to me necessary to use every procedure possible, every trick conceivable. Hence, positive regulatory processes as well as negative ones. Ours was a difference of nature. Jacques wanted to be logical, even purely logical, while he considered me as being mainly intuitive. Which would not have disturbed me if he had not injected into his remarks a bit of irony, even scorn. But it was not enough for him to be logical. Nature also had to be logical, to function according to strict rules. Having once found a "solution" to some "problem," it had to stick to it from then on, to use it to the end. In every case. In every situation. In every living thing. In the last analysis, for Jacques, natural selection had sculpted each organism, each cell, each molecule down to the tiniest detail. To the point of attaining a perfection ultimately indistinguishable from what others recognized as the sign of divine will. Jacques ascribed Cartesianism and elegance to nature. Hence his taste for unique solutions. For my part, I did not find the world so strict and rational. What amazed me was neither its elegance nor its perfection, but rather its condition: that it was as it was and not otherwise. I saw nature as a rather good girl. Generous, but a little dirty. A bit muddle-headed. Working in a hit-or-miss fashion. Doing what she could with what was at hand. Hence, my tendency to foresee the most varied situations. But on this question, Jacques was adamant.

Our article was entitled "Genetic Regulatory Mechanisms in the Synthesis of Proteins." The manuscript was sent to the editor of the *Journal of Molecular Biology* on Christmas Eve 1960. Just ten years after my arrival in the attic!

When I left the laboratory that day, the snow had stopped falling, but the weather was still gray and overcast. It was cold. The sidewalks were covered with a thick white layer of snow. On the street the salt had transformed it to mud. The sounds of steps and cars were muffled. And on this Christmas Eve, the streets were crowded. On the boulevard Montparnasse, the people were floundering through the slush. Running in short steps to make their purchases. The Luxembourg Gardens, however, were deserted. A sumptuous whiteness. Immaculate. Buried in a mantle of silence. With a horizontal light. A blend of gentleness and sadness evoking impressions from my childhood. Another garden in the snow, in Dijon, where, during the Christmas holidays,

I went to play, most often alone. A garden I peopled with bandits, savages, ferocious beasts. Where I felt constantly followed by unseen eyes, observed through branches, watched from behind each trunk, spied on through each hole in the wall. A silence heavy with cold shadows and menace.

It started to snow again in the Luxembourg Gardens. The light grew dim, became tinged a dirty white, then a dark gray. As though one were folding up the day to stow it away in its box. To give way to the night, to haunting memory, to dreams, to terrors. As I was leaving the gardens, I suddenly had an idea for an experiment on cell division. A quite simple experiment. All I had to do was . . .

Index

Index

Index

Index